TIME – The New Definition

Chavan's UED research, Volume 11

Sandeep Chavan

Published by SANDEEP CHAVAN, 2025.

TIME – THE NEW DEFINITION

First edition. June 1, 2025.

ISBN: 979-8230771708

Written by Sandeep Chavan.

Table of Contents

To my beloved father,**Late Shri Jagannath Chavan**,whose quiet strength, wisdom, and life of simplicity continue to guide my every thought.

To my dear children —**Shravani**, **Aradhya**, and **Ninad** —may you always walk in harmony with your inner rhythm, unbound by the illusions of time.

And to all my **students**,past, present, and future —you are not just learners of science,you are seekers of truth.

This book is for you —to remember what time could never teach,but silence always knew.

"Time is not what moves — it's what delays.And the moment you stop waiting, it disappears."

— Sandeep Chavan

Disclaimer

This book presents an original and evolving framework — Universal Energy Dynamics (UED) — developed by the author to reinterpret the nature of time, perception, and reality. While grounded in philosophical reasoning, observational insight, and independent research, it is not intended to replace mainstream scientific models, but to offer an alternative lens for deeper inquiry.

The ideas, terms, and interpretations herein — including concepts like *ATT (Alignment Threshold Time)*, *ripple delay*, and *field misalignment* — are part of a new paradigm and may not align with conventional physics or academic consensus.

Readers are encouraged to approach this material with curiosity, not conformity, and to explore it as a philosophical and experiential proposition, not a final authority.

This work does not provide medical, psychological, or legal advice.

All experiments or applications derived from this material should be conducted with appropriate scientific rigor and personal responsibility.

Opening Note

Before You Count Another Second

You've been told that time flies. That it heals. That it waits for no one. That it's money, it's a dimension, it's an arrow, it's the fourth coordinate of the universe.

And yet—none of it ever felt complete, did it?

Because deep down, you've always suspected that time is not what they told you it is.

You've watched how time stretches when you're waiting, how it vanishes when you're in love, and how it drags during a lecture, only to sprint during holidays. Clocks stay constant. But you don't. Something inside you keeps shifting.

That "something" is not time itself. It's you, adjusting to consequences.

In this book, I'm not here to offer motivational hacks or quantum mysticism. I'm not here to validate time travel fantasies or glorify ancient calendars. I'm here to offer a clean, grounded alternative to what we think time is.

I'm offering a new definition:

Time is not a dimension. Time is a consequence.

It's not the background of existence — it's the aftereffect of interaction. It emerges only when your intent meets resistance, when ripples collide, when fields fail to align instantly.

You don't pass through time. You generate it — every time you act, react, or perceive.

Through this book, we will dismantle the illusions we've inherited: from Newton's absolute ticking to Einstein's bending fabric. We'll leave behind the trap of past-present-future and re-enter reality through the lens of Universal Energy Dynamics — a framework where time is not a container, but a signal. A delay. A consequence.

Each chapter is a conversation — not a lecture. It won't demand equations, only attention. It won't worship time — it will reveal it.

And by the end, I hope you won't just *understand* time differently. You'll *experience* it differently. You may even stop fearing it.

Because once you see the truth — that time doesn't flow, and clocks don't control your life — you'll realize you were never running out of time.

You were just running out of alignment.

Let's begin.

"You don't have time. You have consequences waiting to happen."

— **Sandeep**

How to Read This Book

This book is not bound by conventional chapters, nor is it meant to be read passively from cover to cover in a single sitting.

TIME: The New Definition invites you to *participate*, not just read. To *feel*, not just analyze. To *observe*, not just remember.

Here's how to make the most of it:

1. Come As You Are — No Prior Expertise Needed

You don't need a background in physics, philosophy, or spirituality. The only prerequisite is a willingness to question what you've always assumed about time.

2. Start from the Middle, if You Wish

Each chapter is written to stand on its own — as a lens into one aspect of time, perception, or ripple dynamics. You can begin with a chapter that resonates most, then move forward or backward as you feel drawn.

3. Pause When You Feel Something

This isn't a race to finish. Many of the insights in this book are not just informational — they are experiential. When something shifts in your perception, stop. Reflect. Watch what happens in your own timeline.

4. Use the Appendices as Anchors

The appendices are designed to clarify, connect, and extend key ideas — both scientifically and spiritually. If a chapter feels abstract, the appendices may help you ground it.

5. It's Okay Not to Understand Everything

This is not a textbook. It's a field guide to something you already feel but may not yet have words for. Let your own awareness unfold alongside the material.

6. Return Later — Time Will Read Differently

Come back after a week, a month, or a year. Your understanding will evolve. Why? Because you are the clock — and the book will ripple differently every time you align with it.

You're not here to memorize time. You're here to remember what happens when it disappears.

Prologue: Before Time Began

Imagine this:

You're in a silent, vast space.

No clocks.

No sun.

No tick. No tock.

Just silence.

Stillness. Existence.

Would *time* exist?

Before the first heartbeat, before the first ripple, before light moved or matter formed — was there time?

The honest answer: No. There was only potential — waiting. What we now call "time" didn't exist. It began only when something tried to happen.

That's the radical realization this book begins with.

Time didn't come *before* events. It was never a container, a flow, or a cosmic canvas.

Time is born after something tries to interact.

It is not a fourth dimension. It's a perceived consequence of ripple misalignment — the delay between intent and result. Between cause and response. Between disturbance and equilibrium.

In modern science, time is treated as a measurable entity, a dimension stitched into the very fabric of space. Einstein bent it. Clocks measure it. Calendars worship it. But no one really knows *what* it is. They only know how to count it.

But counting shadows doesn't make them real.

In ancient traditions, time was sacred, cyclical, eternal. In modern society, it's scarce, linear, transactional. But in both — we've mistaken the effect for the thing.

We've mistaken the symptom for the structure.

What if time is not running — you are? What if clocks only track your ripple delays? What if healing, growth, and success aren't products of time — but alignment?

In Universal Energy Dynamics (UED), time doesn't exist independently. It emerges only when ripple fields interact and produce a delay in consequence. That delay is what we've romanticized, feared, and commercialized as "time."

So this book offers a reset — a complete realignment of how we see time. Not as a dimension. Not as a mystery. But as a consequence.

A ripple delay.

And once you understand that, everything else changes:

- **You stop fearing age** — because age is ripple fatigue, not calendar count.
- **You stop rushing** — because urgency is misalignment, not scarcity.
- **You stop chasing the future** — because it doesn't exist. Only ripple paths do.
- **You stop clinging to the past** — because it's not behind you. It's stored within you.

We'll move gently, chapter by chapter, into these ideas. No rush. Because once you see what time really is — you'll realize:

Time never moved. You did.

And that's where our story begins — before time begins.

Part I — Breaking the Clock

Chapter 1: What Is Time, Really?

"What if you've never actually experienced time — only your reaction to something?"

Take a moment.

Close your eyes and try to feel time.

Not your watch. Not the ticking fan. Not your racing thoughts. Just... time.

Can you?

Most people can't. Because there's nothing to feel. You may sense stillness, or movement, or even boredom. But time? It never arrives. It never knocks. It never introduces itself.

It only makes itself known when something happens. A delay. A heartbeat. A disappointment. A surprise. A silence before a message. The gap before the exam results. The wait before the traffic light turns green. We never feel time directly — we feel our response to something not yet resolved.

That's the hidden truth we've missed all along: You're not experiencing time. You're experiencing reaction.

And what we call "time passing" is actually just your mind tracking misalignments — gaps between your expectation and consequence, your desire and delivery, your intent and the universe's response.

This is why 2 minutes in an elevator feels longer than 2 hours at the beach. Why childhood summers seemed to last forever, but adult months vanish in a blink. Why grief makes days feel infinite. And why love makes them disappear.

Clocks stay constant. But you don't. Because clocks track periodic change — you track ripple alignment.

So if time is not what we thought — not a flowing river or a ticking bomb — then what is it?

That's what this chapter is about.

We'll begin by gently taking apart what we've been told — the linear story, the ticking timeline, the arrow of time. Then we'll introduce the deeper truth from within Universal Energy Dynamics (UED):

Time is not a thing. Not a medium. It's a consequence — the ripple delay between intention and alignment.

And once that clicks, your whole perception of reality — and your place within it — will start to shift.

Let's go. You're not running out of time. You're about to meet it — for the very first time.

The Illusion We Live With

Time as Linear, Measurable, Countable

Ask a child what time is, and they'll point to a clock. Ask an adult, and they'll mention deadlines, schedules, and aging. Ask a physicist, and you'll hear about equations, entropy, or space-time curvature.

Across all these layers, there's one common assumption — time is real because we can measure it.

But this is the first illusion.

For centuries, we've treated time as something linear — an arrow stretching from the past, cutting through the present, and vanishing into the future. This timeline is deeply wired into our language:

- "What's next?"
- "Don't dwell on the past."
- "Time will tell."

- "It's only a matter of time."

And so, time becomes a narrative device. A way for us to place events in sequence. A rope tying moments together so life doesn't feel like random static.

But here's the problem: we mistake this sequence for reality itself. We believe the map is the territory.

We assume:

- Because clocks tick, time moves.
- Because events unfold, time flows.
- Because we age, time must be acting on us.

But consider this: Does a clock measure time — or does it just mark repetitive motion?

A pendulum swings. A quartz crystal vibrates. Earth rotates. All these are cyclical processes. They generate regular change — but they don't prove that time exists independently. They simply manifest rhythm. We've attached units to that rhythm — seconds, minutes, hours — and declared it "time."

But it's not time. It's just periodicity.

The clock doesn't tell time. It tells us how many cycles have passed since we started counting.

This is like measuring distance by counting footsteps — then believing footsteps are a dimension. They aren't. They're a way to traverse it. Likewise, seconds aren't time — they're a way we traverse experience.

Now let's zoom in on the second illusion: Time as Duration.

When we say something "took a long time," what do we mean?

Not the number of ticks. We mean:

- The effort felt longer

- The anticipation was unresolved
- The experience was uncomfortable or intense

In reality, "duration" is often a felt sense, not a fact. This is why emotional states distort time. Boredom stretches it. Fear freezes it. Joy dissolves it. That's not because time is elastic. It's because you are the clock — and your state defines the rhythm.

Yet we continue to externalize it.

We say:

- "I wasted time."
- "Time is slipping away."
- "There's not enough time."

But time isn't a resource. It's not a thing you can waste or save. It's not sand running through an hourglass. It's your awareness moving across consequence fields. It only appears scarce when your internal ripple is out of sync with the field around you.

So where did this illusion come from?

The Birth of Linear Time

Much of the modern view of time came from:

- Newton, who imagined time as absolute — like an invisible, universal clock ticking in the background of reality.
- Industrial revolution thinkers, who needed standard time to coordinate trains, work shifts, and global trade.
- Einstein, who bent time into the geometry of space — and while more advanced, still preserved the idea of time as *something* to manipulate.

But all of this treats time as thing-like — as if it's part of the furniture of the cosmos.

In truth, it's not.

Time is not an ingredient of the universe. It's a lens through which we track ripple consequence.

We'll explore this deeply in the next sections, but for now — the key shift is this:

You're not living *in* time. You're living through interactions. And the rhythm, delay, or distortion of those interactions — is what your mind records as "time."

So why are we still measuring it with numbers?

Because numbers give us certainty. Even if the certainty is false.

The Ripple View – Time as Consequence

Time Is Not a Medium — It's a Delay in Response

Let's drop the ticking clock. Let's drop the timeline. Let's even drop the idea that the universe is "moving forward."

Now imagine this instead: A still field. Not a vacuum — but a Structured Vacuum Energy (SVE) field. A perfectly balanced, stable medium — silent, intelligent, undisturbed.

Now place a ripple in that field. A tiny imbalance. A push. A desire. An intent.

What happens next?

That ripple spreads. It tries to resolve the imbalance. It interacts with other ripples. It pushes, pulls, interferes, resonates, or gets blocked. In this dance, something new emerges — consequence.

And the delay between the original ripple and its consequence — that's what we call time.

Time, in the UED framework, is nothing more than a delay in ripple alignment.

The Birth of ATT – Alignment Threshold Time

Let's make it sharper.

In Universal Energy Dynamics (UED), every ripple interaction operates on a unique rhythm. Just like every human heart has its own beat, every energy field — whether it's atomic, emotional, or galactic — has its response interval.

This interval is not fixed in seconds. It's defined by how long it takes for an intent ripple to align with an external field and trigger a response.

We call this the Alignment Threshold Time (ATT).

ATT is not about duration. It's about threshold — the exact moment when a ripple reaches sufficient coherence with its environment to cause something to happen.

For example:

- A thought becomes action when ATT is crossed in the mind-body field.
- A seed becomes a plant when ATT is crossed in the environmental field.
- A decision becomes reality when ATT is crossed in the collective field.

You never act in time. You act in ATT.

Time, then, is just the recorded delay between ripple initiation and its alignment with reality.

Ripple Delay vs. Clock Time

Let's contrast these two models clearly.

Concept	Clock Time	Ripple Time (UED View)
Basis	Repetition (oscillations)	Consequence (alignment threshold)
Nature	Linear, uniform	Variable, field-dependent
Reality Status	Treated as real	Emergent from interaction
Flow	Constant	Only felt when misaligned
Measurability	External, objective	Internal, perceptual
Impact on Experience	Indirect	Direct (felt as resistance/delay)

This table exposes the flaw of standard physics: It counts time, but doesn't feel it.

Only in UED do we feel time — because it arises inside us, from the interaction between our ripple intent and the external field we're engaging with.

You Are the Clock

This is why two people in the same situation don't experience time the same way. In an exam hall, one student finds time flying, another finds it suffocating. Why?

Because their internal ripple intent is interacting with the same environment but producing different ATT delays.

- The confident student: alignment is quick → ATT is short → time feels fast.
- The anxious student: alignment struggles → ATT is long → time feels slow.

The experience of time is not universal. It is a ripple-perception, unique to the observer.

This redefines the observer effect entirely. You don't just observe reality — your internal ripple field dictates your experience of time itself.

And this has profound implications:

- A person with high emotional turbulence will feel time dragging — they're stuck in ripple loops.
- A meditator, in full alignment, feels time dissolve — because no ATT delay is being experienced.
- A hyper-productive person doesn't "manage" time — they align better and reduce ATT friction.

Ripple Harmony Collapses the Need for Time

In a fully aligned system — no delay exists.

A master artist in flow doesn't "lose" time. They enter a state where ATT = zero. Their internal ripple synchronizes so well with the external field that consequence becomes instant.

There is no *time* in harmony. There is only *expression*.

Which means...

Time only exists when there is resistance.

When there's misalignment between your ripple intent and the universal field, you feel time as drag, as delay, as uncertainty, as suffering. But in truth — you're just experiencing the energetic consequence of friction.

So what we call:

- **Stress** → is high ATT friction
- **Urgency** → is ripple conflict
- **Restlessness** → is delayed alignment
- **Waiting** → is perception of misalignment
- **Patience** → is trust in ATT unfolding

Let's summarize this shift:

- Time is not a line you walk on. It's a lag in ripple feedback.
- You don't live in time. You live in ripple interactions.
- You are not running out of time. You are running into misalignment.

This is the ripple view. And once you accept it — time will no longer be a master, but a messenger.

Real-Life Applications

Why 5 Minutes Feels Like Forever (Or Never Happens)

Now that we've stepped away from the ticking clock and into the ripple field, let's bring this down to earth — to daily life. To the strange, intimate ways in which time deceives us.

Because once you understand that time is just ripple delay, everything starts to make eerie sense.

Let's begin with a classic:

The Waiting Room Paradox

You're at a clinic. You're told, "The doctor will see you in 5 minutes."

Ten minutes pass. You feel like it's been thirty. Why?

Because you're not tracking time. You're tracking unmet consequence. Your mind has already cast a ripple — expectation: *"I will be called."* But the external field hasn't responded. So ATT stretches. Misalignment grows. You feel that friction as *slow time*.

Meanwhile, the receptionist, whose ripple is aligned with the process, feels no distortion. Her ATT is minimal. For her, time is smooth.

Now flip the scene:

You're in a movie theater, lost in a powerful story. Ninety minutes vanish. Because you're not waiting for consequence. You're moving with it. Ripple feedback is smooth. The mind is not tracking resistance. It's immersed. So ATT is compressed — and time disappears.

When there is no resistance, time dissolves. When there is misalignment, time distorts.

Productivity and Time Management: The Myth

We've been sold the illusion that productivity is about "managing time." But if time is ripple delay, then you can't manage it like a file or calendar. You can only manage alignment.

This explains why:

- To-do lists often don't get done.
- Multitasking increases stress.
- "Time-blocking" can still feel overwhelming.

The issue is not the schedule — it's your ripple state.

Let's say you plan to write a report from 9 AM to 10 AM. But your intent ripple is fragmented — distracted, unaligned, emotional. You sit for 60 minutes, and barely a page is written.

Now imagine the opposite. You're centered, clear, and fully present. You write 3 pages in 25 minutes — and feel energized.

In both cases, the clock shows an hour. But one was ripple resistance, the other was ripple alignment.

The first "took more time." The second was more real.

Productivity is not about doing more in less time. It's about reducing ATT — the delay between intent and aligned action.

Emotions as Time Distorters

Here's the secret nobody tells you: Your emotional state defines your time perception.

- **Anxiety**: Ripple field is hyperactive, misaligned with external pace. Result? Time stretches.
- **Depression**: Internal ripple is dulled, response is slow. ATT elongates. Days feel endless.
- **Joy**: Ripple field is in harmony, ATT drops. Time compresses. You feel "alive."
- **Grief**: Ripple shock delays feedback integration. Time halts or loops. "It feels like yesterday."

Why does heartbreak make a single day feel like a month? Because ripple consequences keep re-triggering. ATT becomes a dense loop. You're caught in unresolved feedback — hence, *time drags*.

Now consider love — the early stage, where every moment feels electric. You don't notice time. Because you're not waiting. You are experiencing aligned ripple exchange. That's why it feels timeless.

So emotions are not just moods. They are time machines — warping your ATT experience.

Healing and Growth: Not Time-Based, But Consequence-Based

You've heard it: "Give it time." But that's just a polite way of saying: "Let your ripples re-align."

A wound heals not because time passes, but because the biological and emotional fields restore balance. Sometimes that happens in days. Sometimes it never happens — even after years. So it's not the calendar. It's the realignment process.

The same applies to:

- Personal growth
- Trauma recovery

- Forgiveness
- Spiritual awakening

These are not time-bound. They are alignment-bound.

And once you stop watching the clock, and start watching your ripple field, you'll know where you are in the process.

Communication and "Right Timing"

Ever tried to explain something to someone — but they just couldn't "get it"? Then days later, they suddenly say, "Now I understand!"

That's ATT.

The message ripple was sent. But the receiver's field wasn't aligned. So consequence (understanding) was delayed.

What we call "right timing" in communication, relationships, business — is really ATT synchronization. When both ripple fields meet at a resonance point, consequence happens instantly. A decision is made. A deal is closed. An insight lands.

Insight is not a lightning bolt. It is a ripple that finally reached you.

Why Some Lives Feel Longer

Here's a deep one.

Two people may live 70 years. One feels it was over in a flash. The other feels like they lived a thousand lives.

What's the difference?

Not the count of years — but the intensity of ripple consequence. When someone lives with awareness, emotion, depth, and courage — they trigger more ripple responses. More ATT loops. More resolution. More inner movement.

That's why a year of adventure can feel fuller than a decade of routine.

So "long life" isn't about lifespan. It's about consequence density.

Practical Takeaway: Don't Manage Time. Manage Ripples.

If you want to:

- **Save time** → Reduce ATT
- **Be present** → Enter ripple lock
- **Stop wasting time** → Align with consequence
- **Heal** → Wait for ATT resolution
- **Create impact** → Trigger aligned ripple chains
- **Live fully** → Engage ripple-rich fields with awareness

Then time becomes irrelevant. Because what you seek is never *in* time. It's in alignment.

Closing Reflection

"Is it time you feared — or what you were waiting for?" *Time is your consequence radar.*

Pause now, just for a moment.

Look back at your life — the moments you said you didn't have "enough time," or that "time ran out," or that it was "too early" or "too late."

Was it really time that failed you?

Or was it that something you expected didn't happen — and in that gap of consequence, you blamed time?

That's the shift this chapter asks you to make.

Stop thinking of time as a force acting on you. Start seeing it as a field of response — a mirror that reflects how aligned or misaligned your intent is with reality.

When your actions, expectations, and environment resonate — time feels soft, even invisible. When you're disconnected — time shows up, not to punish you, but to signal you:

"Something hasn't landed yet."

That's why time feels heavy in waiting, light in love, brutal in regret, and infinite in awe.

Time is not the enemy. It's the echo. It's the radar ping that tells you: "Your ripple hasn't returned yet. Stay aligned."

It doesn't flow. It responds.

It doesn't pass. It alerts.

And it doesn't run out — you simply miss the moment of alignment.

Perception Prompt:

Take a quiet minute today and ask:

"Where in my life do I feel like I'm running out of time?" Now ask: "What consequence am I really waiting for that hasn't happened yet?"

You'll see — it's not time that's the issue. It's the ripple that hasn't closed.

With this, we close Chapter 1 — not with a full stop, but with a lens:

Every second you feel... is not a tick. It's a message. Time is not measuring you. It's messaging you.

Next, we'll challenge the most cherished belief of all — that time *moves* at all.

Because what if time never moved?

Only *you* did — through fields, intents, resistance, and echoes.

Ripple Delay vs Clock Time, ATT Map

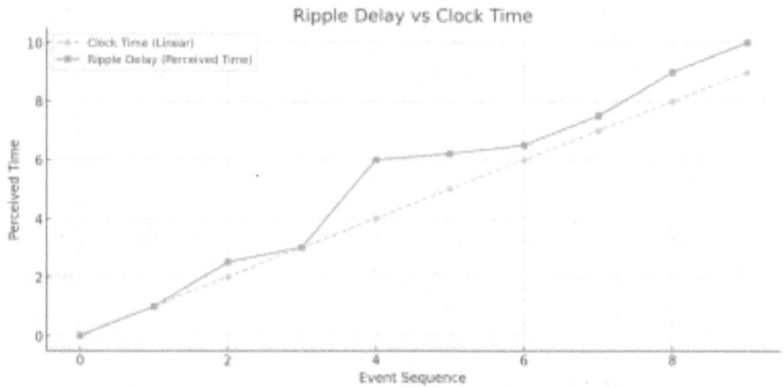

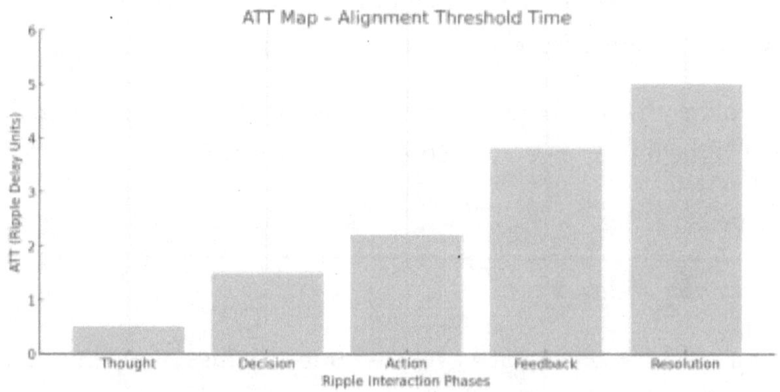

◈ **Ripple Delay vs Clock Time** – showing how ripple-based time perception diverges from linear clock time.

◈ **ATT Map (Alignment Threshold Time)** – showing how different phases of ripple interaction involve varying delay thresholds before consequences emerge.

Chapter 2: Time Never Moved

"If nothing in the universe moves, what's making the clock tick?"

Look around.

Everything seems to be in motion: Planets spinning, stars blazing, cities buzzing, people rushing, seconds slipping by. You check your watch. Another minute gone. But gone where?

We assume motion is everywhere — it defines our reality. And where there's motion, there must be time, right?

But here's the heresy we're about to unpack: What if nothing is actually moving? What if the universe — at its core — is perfectly still?

And if that's true...

What, then, is making the clock tick?

A clock is just a machine repeating a cycle. It counts swings of a pendulum, vibrations of a crystal, or pulses of atomic resonance. But these are not motions through time. They are ripple realignments within a structured field.

Ticking doesn't prove that time is moving. It only proves that something in the field is oscillating — resetting its own misalignment over and over.

That's it. Tick... realignment. Tock... realignment. Repeat.

So what if the ticking clock isn't tracking a moving universe — but measuring its own inability to stay in alignment?

Let's take it further. If the whole universe is a field of ripples — and those ripples arise only when disturbed — then in the absence of disturbance, there is no ripple, and therefore no motion, and therefore no time.

Stillness is the default. Motion is the ripple. Time isn't the backdrop — it's the tremor after the push.

This flips everything.

- We don't move through time.
- We create the illusion of motion by observing changing ripple patterns.
- Our sense of movement is constructed through perception — not physical translation.

So this chapter isn't just about clocks. It's about motion itself — what it really is, and how time became its ghostly twin.

We'll revisit Newton, gently challenge Einstein, and ultimately reset the idea of motion as a change in alignment, not an actual shift through space.

And we'll begin to see:

Time never moved. You did — through layers of resistance and return.

Ready to stay still while reality unfolds?

Let's begin.

The Illusion of Motion and Flow

Why Everything That Moves May Actually Be Standing Still

Since childhood, we've been taught to believe in movement.

- The Earth spins at about 1,670 km/h.
- It orbits the Sun at 108,000 km/h.
- The Sun itself orbits the Milky Way at over 828,000 km/h.
- And somewhere in that cosmic chaos, you're told to sit still and breathe.

It's dazzling. Terrifying, even. But here's a question rarely asked:

If we're moving at such impossible speeds, why do we feel... still?

Why don't we feel *anything*? Because what we call "motion" might be an illusion of perception, not a fundamental truth.

Motion: A Story We Tell Ourselves

Let's define motion in its conventional sense:

The change of position of an object with respect to a reference point over time.

It sounds objective. But look closely — this definition is observer-bound.

You can't define motion without:

1. A chosen reference frame
2. A sequence of changes
3. A way to measure delay (i.e., time)

This means motion only exists when you observe it. It's not absolute. It's relational. Just like time.

When you're sitting in a train, and another train moves beside you — you feel like *you're* moving. Until the platform appears, and reality reorients. This is not just optical trickery. It's a glimpse into a deeper truth:

Motion is not happening. It is perceived.

Nothing Moves — Everything Aligns Differently

Now bring in Universal Energy Dynamics (UED).

UED proposes that the universe is fundamentally a Structured Vacuum Energy field — pure stillness. From this field, ripples arise when imbalance occurs.

What looks like "motion" is not an object shifting positions — but a change in ripple alignment between observer and field.

Let's unpack that.

Think of a fan. When it spins slowly, you see the blades. Speed it up, and you see a blur. Speed it up even more — the blades disappear. Now imagine the fan never actually spins — it simply reorients how the field aligns with your perception.

That's how UED sees the universe.

- The Sun doesn't rise — the ripple between you and the light source shifts.
- The Earth doesn't orbit — our perception locks onto ripple harmonics that cycle predictably.
- Nothing "moves" from A to B — the observer's ripple reattaches to different nodes in the field.

It's not motion — it's ripple reassociation.

"You didn't walk across the room. Your ripple shifted its lock point across the field — and consequence emerged."

This breaks the backbone of Newtonian intuition.

Einstein's Relativity — Still Bound to Motion

Even in Einstein's universe, motion persists — just more creatively.

He taught us that time and space are flexible. That gravity bends spacetime. That motion affects time's flow.

And while brilliant, this model still assumes that objects move through a continuum. It doesn't question the premise of motion itself — only its consequences.

But if UED is correct, and the field is all there is, then spacetime isn't bending — it's ripple interaction zones shifting under different field tensions.

So:

- Gravity isn't a curvature — it's a tension field caused by condensed ripple concentration.
- Time dilation isn't stretching time — it's a delay in ripple consequence due to field density.

In UED, motion disappears. Only alignment and delay remain.

Where Does Flow Come From Then?

"But I feel movement!" — of course you do. Because you're not still — you're experiencing a stream of ripple responses. Your brain, your senses, your emotional radar — all track changes in alignment and record them in sequence.

This gives rise to the sensation of flow — a constructed continuity. But beneath that:

- The universe isn't flowing.
- It's echoing — shifting alignment, field by field.

That's why:

- Time feels slow when you're stuck — no new alignment, so no ripple evolution.
- Time feels fast in joy — smooth, consistent realignments.
- Flow states dissolve awareness of time — ATT approaches zero. There's no resistance.

You're not moving. You're resonating.

The Stillness Behind All Change

All motion arises from one hidden stillness — the background field.

Like a movie projector, it never moves — it just displays changing frames. Like a pond, it is undisturbed until touched — and only then do ripples form. Like consciousness, it is present — even when nothing happens.

This stillness is not empty. It is charged with possibility.

And only when intent — your ripple — strikes it, does time emerge. Not because you moved, but because you tried.

Let's recap this illusion:

- The universe is not a flowing river.
- It is a ripple field — still and structured.
- Motion is not travel. It's relational reassociation between your ripple and the field.
- Time is not flow. It's the gap between attempted alignment and actual consequence.

So when you say something "moved you" — it did. Not across time. But across layers of resistance within a still, intelligent field.

The world didn't turn. You did.

And now that we've shattered the illusion of motion, let's move forward — or rather, align deeper — into what clocks are really measuring... and why they deceive even the sharpest minds.

Next: The Machine That Can't Stop Ticking

The Ripple Clock – Why Clocks Tick in Stillness

Machines Move, but They Don't Know Why

Let's talk about clocks — the instruments we trust to *tell us the truth* about time.

They tick with precision. They measure in equal intervals. They promise objectivity. But here's the truth through the lens of Universal Energy Dynamics (UED):

A clock is just a powered ripple machine — it runs without knowing what it measures.

Motion Requires Energy — But Energy Alone Doesn't Prove Time

Every form of motion in the physical world — from rotating gears to the orbit of satellites — demands energy. This is true for all machines, including clocks.

A pendulum clock must be wound. A quartz watch runs on electricity. An atomic clock relies on oscillations driven by controlled energy pulses.

None of these machines function without energy input. And all of them perform periodic motion. But what are they doing?

They are not measuring time itself. They are repeating a cycle of ripple rearrangements — oscillations — and counting them. That's it.

The moment power ceases, the illusion collapses. Time, it seems, has "stopped."

But has time really stopped, or has the machine simply run out of ripple energy to sustain its alignment behavior?

Clocks Are Just Consequence Loops

Let's break this down.

A machine doesn't *understand* time. It only:

1. Responds to a power input (internal ripple disturbance).
2. Follows a fixed path of realignment (mechanical or electronic oscillation).
3. Resets its own disturbance repeatedly — creating a predictable ripple loop.

This is what you see as "ticks." What you hear as "tock." What you assign as "seconds."

But from the UED standpoint, all this activity is closed-loop ripple mechanics — an artificial dance within a local field. The machine is not accessing true ripple consequence. It has no awareness of alignment with external fields.

So while it moves, it doesn't know.

Clocks don't measure time. They record how long they've been disturbing themselves.

The Human Error: Interpreting Machine Cycles as Time

Humans made a subtle but critical mistake: We took the output of machine-based oscillations and declared it reality.

We told ourselves: "This many vibrations = 1 second." We made laws, schedules, and scientific equations based on this artificial rhythm.

But time didn't exist in that second. Only the machine's sustained ripple activity did.

It's like calling the sound of a metronome "music." The sound is real. The rhythm is measurable. But music only happens when a mind aligns with it.

Similarly, time is not born in the tick — it's born when awareness interacts with consequence.

The Universe Has No Clocks — Only Fields

Zoom out.

Nowhere in the cosmos is there a device ticking. Stars don't tick. Galaxies don't beep. Black holes don't buzz. Yet everything unfolds with pattern, rhythm, precision.

Why?

Because the universe doesn't use clocks. It uses field intelligence. Consequence alignment. Ripple memory.

- Planets don't orbit *in time*. They maintain field tension and balance.
- Seasons don't arrive *on schedule*. They emerge when systemic ripple conditions realign.
- Growth, aging, decay — all happen based on consequence thresholds, not calendar markers.

The structured vacuum energy (SVE) field does not operate on seconds. It operates on resonance — cause-effect tension that unfolds naturally when the ripple state is satisfied.

That's why, in UED, we say:

"Time is not a quantity. It's a consequence lag — a ripple delay between intention and resolution."

Clocks Keep Going Even When Nothing Happens

Here's the final paradox.

A clock will keep ticking in a sealed room, during a war, or on a dead planet. It doesn't know if:

- Anyone is watching
- Any event is unfolding
- Any consequence is being born

Because a clock isn't aware of the world — it just burns energy to sustain its own loop.

In this sense, a clock is a blind engine of self-realignment. It keeps resetting its own ripple — with zero awareness of external ripple fields.

This is what makes it mechanically precise, but ontologically empty.

Compare that to you — a conscious ripple field. You don't just tick. You perceive alignment or misalignment. You don't just count — you wait, you feel, you respond.

Which is why:

Only a conscious ripple field can truly experience time. A machine can never know time — only simulate its rhythm.

In the next section, we'll dive into what this means for our own experience of time — when time feels like a flood, when it feels stuck, and how alignment compresses or expands it.

We'll explore: Why Flow Feels Timeless — and Stress Makes Time Expand

Ready to experience your own ATT curve in action?

Let's continue.

Applications – When Time Disappears, Distorts, or Freezes

Stillness, Flow, Trauma, and Decision-Making through the Ripple Lens

If time never moves — if it only emerges from ripple misalignment — then the way we *feel* time must also depend on our internal ripple state.

This means your experience of time is not just about your schedule or the clock on the wall. It's about your ripple condition, your field engagement, and how aligned you are with your environmental feedback loops.

Let's explore some profound and practical examples.

1. Stillness in Meditation: When Time Ceases to Exist

People often describe meditation as "timeless." They sit for what feels like five minutes, only to realize an hour has passed. Or, conversely, they sit for 10 minutes and it feels like eternity.

Why?

In UED terms, meditation is a practice of ripple stabilization. You withdraw energy from the outside world — no external stimulus, no emotional triggers, no tasks. You enter ripple neutrality.

When your internal ripple is in perfect balance — i.e., no ripple conflict, no misalignment — there is no consequence delay to track. Therefore:

No ATT = No Time Perception

The more deeply you enter that stillness, the more time "vanishes." You're not in a trance. You're in perfect ripple resonance with the field.

This is not mystical. It is mechanical — an energetic condition of zero consequence resistance.

2. Time Dilation in Flow States: Aligned Ripple = Compressed ATT

Flow — the state athletes, artists, and innovators describe as being "in the zone" — is another form of ripple alignment.

During flow:

- The mind is fully engaged.
- The action and awareness merge.
- Feedback is immediate.
- There is minimal self-reflection or anticipation.

This is a near-zero ATT zone. Every intent meets environment with minimal resistance. There's no waiting, no friction, no looped consequence. Just pure alignment.

As a result:

Time compresses. You are not aware of seconds. You are awareness itself.

In UED, we would say: "Your ripple is surfing the field without distortion."

The field isn't moving — but your alignment with it is so fluid that it feels like motionlessness inside a moving moment.

3. Good Decisions Seem "Timeless" — Why Wisdom Doesn't Rush

Ever noticed how the best decisions — the wisest, clearest, most effective ones — don't seem rushed?

They often arise in a moment of calm clarity. You don't feel pressured, or short on time. You're not calculating — you're *knowing*.

Why?

Because the best decisions emerge when:

- Your ripple field is stabilized
- Your internal intent is clear
- Your perceptual alignment with the consequences is mature

In such a state, your ATT is optimized. You're not reacting to ripple conflict. You're responding in alignment with the wider field. Time does not pressure you — because you are not resisting consequence.

In fact, the most transformative decisions often feel like they *happen to you*. Not because time provided an answer, but because you exited misalignment long enough for clarity to emerge.

"You don't find the right moment. You become aligned enough for the moment to find you."

4. Trauma: When Time Freezes and Echoes

Now let's go deeper — into the darkest time distortion: trauma.

A traumatic event often leaves people saying:

- "It felt like it lasted forever."
- "I'm still stuck in that moment."
- "I relive it again and again."
- "Time stopped."

Trauma freezes ripple alignment at the moment of energetic shock. The mind gets locked in a high-ATT loop — meaning the ripple disturbance never fully resolves.

Instead of the ripple dispersing and integrating, it keeps echoing back — like a stuck wave in a closed chamber.

This ripple feedback manifests as:

- Flashbacks
- Anxiety triggers
- Repetitive emotional responses
- Feeling that "time never moved forward since then"

You are not stuck in the past. You are stuck in an unresolved ripple field — a delay that hasn't realigned.

"Trauma is not a memory. It's a ripple that hasn't returned home."

Healing, then, is not about waiting for time to pass. It's about managing the ripple environment enough to allow dissipation, integration, and eventual re-alignment with the present field.

5. Long-Term Vision: Slowing Perception to See Beyond

Finally, let's talk about long-term thinkers.

They often appear calm in chaos. Patient. Grounded. They don't panic over short-term outcomes. Why?

Because their ripple intent isn't fixated on immediate feedback. They operate from a long ATT awareness — they're attuned to consequences far down the field.

This reduces inner ripple turbulence. It creates composure. And ironically, it accelerates decision quality because they're not chasing short-term alignments that generate frequent misfires.

"The more you trust long-range ripple return, the less you're shaken by the local clock."

In a UED framework, visionary thinking is extended ripple planning, not extended time planning. It's a perceptual stance — not a scheduling strategy.

Summary: Time Emerges from Ripple Behavior — Not the Clock

Let's simplify:

Phenomenon	UED Interpretation
Meditation	No ripple disturbance → No perceived delay
Flow	Instant ripple feedback → Compressed ATT
Wisdom	Clear alignment → Time irrelevant
Trauma	Ripple shock loop → High ATT → Frozen time
Vision	Slow feedback cycle → Broad ripple awareness

You don't need to manage time in any of these cases. You need to observe ripple alignment, ATT behavior, and consequence thresholds.

Once you do — time becomes flexible. Not because it bends — but because your relationship to resistance changes.

Closing Reflection

"Time never moved. You did — through layers of resistance and reaction."

Take a breath. Not just to pause — but to *feel.*

Because in this very moment, you might sense it:

Nothing around you is really moving. Your body may shift. The wind may blow. The screen may glow. But beneath it all, the field remains still.

What you experience as "change" — as motion, flow, progress, loss — is not the world moving through time. It is you, shifting from one ripple alignment to another.

You call it a day. A phase. A journey. But it's all just a series of resonance events — responses to your own energetic state.

Time didn't carry you anywhere. It never had legs. It never flowed. It never pulled you forward or pushed you back.

You moved.

You moved when:

- You reacted emotionally to an unresolved event.
- You waited restlessly for an outcome that hadn't matured.
- You reached alignment and felt peace — as if the whole world paused with you.

These are not experiences inside time. They are time-like feelings produced by ripple activity — or its absence.

Time is not a force. It is a friction. It emerges when ripple intent meets field resistance.

And every time you struggled, doubted, or hesitated — you weren't stuck in time. You were simply cycling through layers of unaligned consequence, waiting for clarity.

So when you say:

- "I've come a long way."
- "I've grown over time."
- "I left the past behind."

Pause and rephrase:

"I've shifted across ripple fields. I've reduced resistance. I've dissolved delay. I've realigned."

Because that's the real movement. That's the real progress.

And that's all time ever was: A consequence radar tracking your journey through misalignment.

Perception Prompt:

Ask yourself: "Where in my life do I feel like something should have 'already happened'?"

Now ask: "What layer of resistance am I still carrying? What reaction am I still stuck in?"

Then ask one final question:

"If time never moved... what am I actually waiting for?"

In this stillness, the illusion fades.

You are not behind. You are not ahead. You are not running out of time. You are running into alignment — one ripple at a time.

And that changes everything.

In the next chapter, we'll go deeper into the illusions of past and future, and why these concepts may be the greatest ripple echoes ever misunderstood.

Ripple Reality vs Mainstream Time: A Comparative Grid

Concept	Mainstream View	UED (Ripple Reality) View
Time Perception	Linear, ticking flow	Ripple delay caused by misalignment
Motion Source	Physical movement through space	Ripple realignment within a still field
Experience of Delay	Duration measured in seconds/minutes	Consequences emerging over ATT (Alignment Threshold Time)
Progress Indicator	Distance covered over time	Shifting across layers of resistance
Reference Frame	External (space-time coordinates)	Internal ripple interaction with external field
Reaction Feedback	Passive; time flows regardless of intent	Active; delay only occurs when ripple intent meets resistance
Perceived Stillness	Rarely noticed unless paused	Fundamental — stillness is the ground state
Emotional Impact	Secondary to logic	Primary determinant of time stretch/compression
Clock Dependency	Essential for measuring reality	Clocks simulate rhythm, not real time

Chapter 3: The Day Einstein Made Time Bend

Respect the revolution. Then reveal the ripple.

"The day Einstein made time bend, the universe applauded — but no one asked: what if time wasn't real to begin with?"

It was one of the greatest plot twists in science. Newton gave us absolute time — steady, silent, ticking across the heavens like God's pocket watch. But then came Einstein, the quiet disruptor with wild hair and a wilder idea:

What if time wasn't fixed? What if it bends, stretches, slows down — depending on how fast you move or how close you stand to gravity?

The scientific world gasped. And rightly so.

For centuries, time had been the invisible ruler of all things — firm, linear, absolute. But with a stroke of theoretical genius, Einstein replaced that ruler with something softer, curved, and elastic: space-time — a unified fabric, where time could be bent like space, manipulated by motion and mass.

It was elegant. It was disruptive. It was revolutionary.

And yet...

What exactly did Einstein bend?

Was it time itself that yielded? Or was it merely our belief in how time behaves?

Here's the twist: Einstein didn't destroy the illusion — he upgraded it.

He set us free from Newton's rigidity, but kept us locked inside a new geometry. Instead of challenging the existence of time, he redefined it as a flexible fabric — still real, still measurable, still essential to the architecture of reality.

In doing so, he created a more sophisticated illusion — but an illusion nonetheless.

Because what if time never flowed to begin with? What if clocks don't track a cosmic current, but count ripple delays in a still universe? What if nothing bends — and all that truly changes is how long alignment takes?

This chapter is about giving Einstein his due — and then going deeper. Because the day Einstein bent time was historic. But the day you realize time never moved at all — that could be transformational.

Let's step into that ripple.

What Einstein Actually Did

Relativity — The Genius and the Geometry

Before Einstein, the universe ran like clockwork — literally.

In Newton's world, time was absolute. It ticked independently of everything, like a divine metronome beating steadily across all space, for all observers, at all times. Events happened *in* time. Time was the backdrop — firm, universal, and unchangeable.

That vision lasted nearly 200 years.

Then in 1905, a young patent clerk named Albert Einstein published a paper that quietly detonated the foundation of physics. In it, he proposed something radical:

The rate at which time flows isn't the same for everyone. It depends on how fast you're moving.

And just like that, the first chisel struck Newton's sacred clock.

Special Relativity: Time Doesn't Flow Equally for All

Einstein's theory of Special Relativity introduced two revolutionary postulates:

1. The laws of physics are the same in all inertial (non-accelerating) frames.
2. The speed of light is constant — no matter how fast you're moving.

To preserve these two ideas, something had to give — and what gave... was time.

Einstein showed that if you moved faster, time slowed down relative to a stationary observer. This wasn't a metaphor. It meant that two observers experiencing the same events could disagree on how much time had passed — and both be correct.

General Relativity: Gravity Bends Time Too

A few years later, Einstein went even further with General Relativity. This time, he took on gravity.

Traditionally, gravity was seen as a pulling force — mass attracting mass. But Einstein reimagined it as a curvature of space-time.

Massive objects like stars and planets don't just pull — they warp the very geometry of space and time around them. This means that:

- A clock placed near a massive object ticks slower than one further away.
- Time literally dilates in a gravitational field.

This became known as gravitational time dilation.

Relativity in Action: Mountain Clocks, Twin Paradox, and GPS

Einstein's theories may sound abstract, but they're very real — and measurable.

1. Clocks at altitude tick faster. A clock on a mountain (farther from Earth's gravity) runs slightly faster than one at sea level. This has been confirmed with atomic clocks.

2. The Twin Paradox. One twin travels near the speed of light and returns younger than the twin who stayed on Earth. This is a direct consequence of velocity-based time dilation.

3. GPS Satellites. Without correcting for both Special and General Relativity, GPS timing would drift by several kilometers per day. The clocks on satellites (in weaker gravity and higher velocity) tick differently than those on Earth.

These aren't just theories — they're *functioning systems* that depend on Einstein's insight. And they all confirm the same truth:

Time is not absolute. It bends. It stretches. It depends.

The Genius of Einstein

Einstein didn't just give us new math. He gave us a new intuition.

He shattered the rigidity of Newton's cosmos and replaced it with relativity — a worldview where time is not universal, but personalized by perspective. Where mass and motion reshape reality itself. Where space and time become one elastic fabric: space-time.

It was revolutionary. Humbling. Beautiful.

He didn't just make time bend — he made our understanding bend. He showed us that time is not fixed by the heavens, but shaped by experience.

And in doing so, he became one of the greatest scientific minds in history.

But... What Did He Actually Bend?

Here's where we gently challenge the frame.

Einstein proved that clocks run differently in different conditions. But he never asked: *what exactly is the clock tracking?*

Is it time itself? Or is it something else — something deeper?

Because if clocks slow down in a field, and that slowing can be measured, what we're really observing is delay in response — a ripple distortion in consequence, not the deformation of a dimension.

What if...

- Time isn't bending.
- Clocks aren't revealing a flow.
- And space-time isn't a fabric, but a field tension map — showing how ripples struggle or ease their return?

What if time never moved at all — and all that changed was how long it took consequences to align?

That's the threshold where UED begins.

Let's cross it.

The Hidden Assumption in Relativity

Why Geometry Isn't the Answer

Einstein's brilliance lies in what he challenged — and also, in what he preserved.

He shattered the old dogma of absolute time, unshackling physics from Newton's rigid clockwork. He introduced a fluid, dynamic reality where your velocity or gravity environment changes how you experience time.

And yet — for all his audacity, Einstein never questioned whether time itself was real. He simply reshaped it.

That was his hidden assumption.

He liberated time from uniformity — but then made it bendable, spatial, and curved inside a mathematical construct known as space-time.

He freed time from Newton's prison — and then rehoused it in a flexible cage made of geometry.

Time as Geometry: The Quiet Concession

At the core of both Special and General Relativity is the idea that time is not just something you measure — it's something that has shape. Along with the three spatial dimensions, time forms the fourth: a temporal axis. The four together form a continuum — a geometric structure where positions and events are plotted like coordinates.

In this space-time continuum:

- Objects follow geodesics (the shortest path through curved space-time).
- Mass bends space, and time bends with it.
- Motion slows time, gravity warps it — because geometry is dynamic.

To make the math work, time is mathematically represented as a dimension — an axis perpendicular to the other three, with its own distortions based on energy conditions.

But in doing so, Einstein made a critical move:

He *assumed* that time is a real, fundamental structure — just like space — but with strange elastic properties.

This was more than mathematical. It was conceptual framing.

He didn't ask: *What if time isn't a structure at all?* He asked: *If time were a structure, how would it behave under force and motion?*

Causality as Path, Not Process

In Einstein's model, causality — the relationship between cause and effect — becomes geometry-dependent.

An event A causes event B if B lies within A's light cone — the area of space-time reachable by light or signals traveling below light-speed.

This is brilliant as a framework — but it's still spatialized thinking. It assumes:

- That events happen in spacetime
- That causes and effects are mapped like pins on a curved grid
- That the geometry of the grid determines how time and causality behave

But this frames reality as a mathematical terrain, rather than a consequence field.

It still needs motion. It still needs distance. It still needs clocks.

And here's the problem: geometry can only stretch so far before it breaks.

UED Counterpoint: Time is Not a Geometric Axis — It's Ripple Delay

Universal Energy Dynamics (UED) offers a stark alternative.

There is no space-time fabric. There is only a structured field — SVE (Structured Vacuum Energy) — in which ripples arise due to imbalance, intention, or disturbance.

These ripples do not travel through time. They interact with resistance in the field. And the delay between ripple initiation and ripple response is what we *feel* as time.

Time is not a dimension. Time is consequence lag — the byproduct of trying to realign with a resisting field.

In this model:

- A clock doesn't tick because time flows.
- It ticks because it resets its ripple loop under consistent energy input.

- What changes under gravity or speed is not "time" — it's the delay in ripple resolution.

Let's say a ripple in a dense field returns slowly. Einstein would say "time has slowed down." UED says: "Field resistance has increased. ATT has stretched."

No geometry. No warping. Just energy, resistance, and feedback.

Time doesn't curve. It emerges from how hard it is to complete a ripple interaction.

The Einstein Frame Was Not Wrong — It Was Incomplete

Einstein wasn't incorrect. His framework works — mathematically, observationally, practically.

It accounts for real phenomena:

- Gravitational lensing
- Time dilation in orbit
- Expansion of the cosmos

But it only works because our machines are tuned to geometry — and not yet aware of field resistance as a primary factor.

Einstein measured what could be observed — delay, frequency shift, curvature. But UED asks: *What's underneath that?*

What if the curvature is an effect, not a cause? What if time isn't being bent — but our ability to complete consequences is being distorted?

What if the very act of defining time spatially has blinded us to its true source?

"Einstein liberated time from being absolute — but kept it hostage as a bending fabric."

It's time to take time out of space — and place it back into consequence mechanics.

UED Response — No Fabric, No Flow, Just Ripple Delay

Where Universal Energy Dynamics Parts Ways with Relativity

Einstein's revolution showed us that time bends — but UED suggests a more fundamental revelation: Time never bent. Time never moved. It only emerged as resistance in a ripple field.

To understand this shift, we need to leave behind the illusion of moving objects and curved coordinates. Instead, we enter the world of Structured Vacuum Energy (SVE) — the foundational field in UED, where nothing ever truly moves.

Instead of motion, what we perceive as change, causality, gravity, and time — are all just ripple realignments within this field.

The Core UED Claims (Reframing the Foundation)

Let's clearly articulate where UED diverges from relativity:

1. **Nothing moves.** What appears as motion is a sequence of ripple alignments. You do not move through space. You re-align with different field points across the ripple structure.
2. **All events are ripple realignments.** What we call an "event" is not a point in spacetime, but a ripple interaction. Every consequence arises when a disturbance reaches a threshold of alignment with the field.
3. **Gravity is not curvature.** UED posits gravity as ripple tension density. Heavier mass → more concentrated ripple field → higher resistance → slower realignment. There is no geometric warping — only field compression and tension buildup.
4. **Time is not bent.** What Einstein interpreted as time dilation is simply the lag in ripple response due to increased field resistance.

The delay in feedback is not a property of "time" — it's a function of ATT (Alignment Threshold Time).

So, UED doesn't discard Einstein's observations — it reinterprets them. Where he saw bending space-time, UED sees ripple delays in a still, intelligent field.

The Falling Apple — Not Gravity, Not Curvature

Let's revisit Newton's famous apple.

Einstein would say: the apple falls not because Earth pulls it, but because space-time around Earth is curved — and the apple follows a geodesic path through this geometry.

UED says: the apple falls because its ripple field is imbalanced in the presence of Earth's denser ripple structure.

The Structured Vacuum beneath the apple is not empty — it's a field of potential alignment. The apple's ripple seeks equilibrium. As it aligns with the denser Earth field, it realigns downward, appearing to fall.

Gravity, then, is not a curvature in fabric — but a response to ripple pressure in the field.

No curvature. No warping. Just energy rebalancing.

Redefining Gravitational Time Dilation — It's Not Time That Slows

In Einstein's model, a clock runs slower closer to a massive object because space-time is stretched by gravity.

UED offers a different interpretation:

- A clock placed in a denser ripple field (closer to Earth's core, for example) experiences greater ripple resistance.
- Every cycle of the clock (e.g., the vibration of an atomic particle) takes longer to complete — not because time is distorted, but because ripple feedback takes longer to realign.

So the delay isn't happening *in* time. The delay is the experience of time.

This is ATT in action — the Alignment Threshold Time required for a ripple system to reach response. In denser fields, ATT increases. Hence, slower clocks.

Einstein: Time slows in gravity.

UED: Ripple realignment takes longer in dense fields.

No space-time stretching. Just delayed consequences.

Analogy: The Road Didn't Bend — You Took Longer to Hear the Echo

Imagine shouting into a canyon.

If the air is dense, humid, or layered with temperature variations, your echo might take longer to return. The geometry of the canyon hasn't changed — but your experience of response has.

That's time, in UED terms.

Time is not a bent path. It is a delayed echo in a resistant field.

The road didn't bend. You just took longer to reach the echo.

This analogy preserves all observable truths of relativity (slower signals, delayed light, frequency shifts) — but it roots them in field resistance, not geometric deformation.

Summary: Relativity Measured the Delay — UED Explains Its Origin

Einstein built a geometry to track when and where things *happen*. UED discards geometry altogether — it only cares *why* things happen *when* they do.

- Relativity calculates the delay based on distance, mass, and speed.
- UED explains the delay as a ripple response resistance — nothing moves, nothing bends, everything responds.

It's not a new equation — it's a new cause.

Time is not flexible geometry. Time is how long it takes the universe to echo your ripple back.

And the more tension your ripple faces, the longer the response. That's not dilation. That's delay.

That's not bending. That's resistance.

And that's where UED completely parts ways with Einstein.

Respect and Realignment — Einstein Was Right for His Time

But It's Time to Go Beyond Time

Before we close this chapter, let's pause — not to correct Einstein, but to honor him.

Because without Einstein, we would still be trapped in Newton's rigid timeline — where everything ticked forward in perfect mechanical sequence, and space was nothing more than a container for objects to fly through.

Einstein gave the universe texture. He let time stretch and bend. He let space respond to energy. He told us that clocks don't tick in isolation — they listen to the cosmos, and beat to its changing rhythms.

Thanks to his insights, we have:

- GPS systems that compensate for gravitational time shifts.
- Satellite technologies and orbital physics that accurately account for time dilation.
- Black hole models that predict event horizons and gravitational lensing.
- A profound understanding that your perspective defines your experience of time.

For all of this, Einstein was right — for his time.

But the Universe Is Speaking Louder Now

And now, we see cracks in the beautiful fabric he designed.

- Galactic rotation curves defy Newton and Einstein alike — stars move too fast for visible matter to hold them. So we patch the theory with invisible glue: *dark matter.*
- Black hole paradoxes grow more severe — with event horizons creating contradictions in entropy, information, and causality. We invent *information loss*, *firewalls*, and *Hawking radiation* to plug holes in understanding.
- Quantum and relativistic models refuse to unite — the smooth geometry of space-time cannot coexist with the wild, discrete behavior of quantum fields.

The deeper we look, the more Einstein's framework becomes a map — not the terrain.

It's still useful. It still describes. But it no longer explains.

UED Doesn't Fight Einstein — It Finishes the Conversation

Universal Energy Dynamics doesn't discard relativity. It completes it.

Where Einstein bent time to explain delayed consequences, UED goes further:

"Einstein bent time to explain consequence. UED removes time — and explains consequence directly."

UED sees the delays not as distortions in a dimension, but as resistance in the ripple field. It doesn't simulate the delay through coordinates. It models it through ripple tension and ATT.

In this light, time doesn't flow. It doesn't stretch. It doesn't bend.

It simply appears when ripple intent meets resistance — and vanishes when alignment is complete.

That's the picture Einstein began to draw. UED simply shows the invisible hand holding the brush.

In the next chapter, we'll explore the illusion that keeps most people tied to the past — believing it's behind them, influencing them, defining them.

But what if the past isn't behind you at all?

Closing Reflection

"Time didn't bend. Our belief did."

Einstein didn't break time — he bent our understanding of it. He loosened its grip, reshaped its frame, and invited us to imagine it as something alive — elastic, reactive, personal. For that, the world should be grateful.

But now, over a century later, we must ask:

Did time itself bend? Or did we simply adjust our belief to match a deeper illusion?

What if the clocks never measured a flowing dimension — but simply tracked how long it took the universe to respond?

What if gravity wasn't pulling, and space wasn't curving — but ripple fields were just realigning under pressure?

In Universal Energy Dynamics, nothing moves. Nothing bends. Everything resonates, resists, and realigns.

Space, gravity, and motion are not geometric stories. They're energetic consequences.

So we must now ask, with renewed awareness: *What else have we geometrized that was never spatial to begin with?*

Maybe time was just the first illusion we wrapped in coordinates. Maybe the map has gotten so sophisticated that we've forgotten we were drawing over ripples all along.

And maybe — just maybe — the past isn't behind us.

Because what lies behind you isn't a line of dates or coordinates. It's a residual ripple field — still echoing, still unresolved, still influencing your present.

And that's where we go next.

Relativity vs UED: Two Views on Time, Motion, and Gravity

Concept	Einstein's Relativity	Universal Energy Dynamics (UED)
Nature of Time	A flexible dimension; part of 4D space-time fabric	Not a dimension; an emergent ripple delay due to field resistance
Time Dilation	Time slows near mass or at high speed (gravitational/velocity)	ATT (Alignment Threshold Time) increases in high-density ripple fields
Cause of Motion	Objects move along curved paths (geodesics)	No true motion; apparent movement is ripple realignment across still field
Space-Time	A real, dynamic fabric that bends with mass and energy	An illusion; no fabric, only SVE (Structured Vacuum Energy) and ripple fields
Gravity	Curvature of space-time due to mass	Ripple tension density; mass is a condensed ripple field causing resistance
Event Reference	Coordinates in a 4D continuum	Consequence in a ripple response field
Clocks & Measurement	Clocks tick slower due to curvature effects	Clocks delay due to resistance in ripple feedback
Causality	Determined by light-cone structure in curved space-time	Determined by ripple interaction thresholds and alignment cycles
Philosophical Frame	Geometry governs physics	Consequence governs emergence
Limitations	Breaks down near singularities; incompatible with quantum field theory	Unified with ripple-based emergence across scales

"Ripple Resistance vs Time Dilation (ATT Delay)".

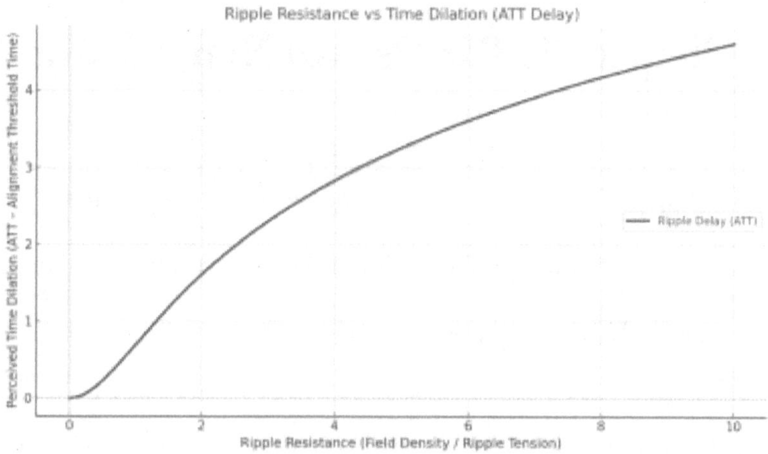

It illustrates the UED concept that as ripple resistance (due to field density or tension) increases, the perceived time dilation also increases — not because time is bending, but because **ripple realignment takes longer** (higher ATT).

"Time doesn't stretch — your ripple just takes longer to return."

Chapter 4: Past Is Not Behind You

A conversational paradox — your past isn't gone, it's ripple residue.

"If the past is really behind you, why does it keep showing up inside you?"

They say time heals. They say the past is over. They say, *"Move on. Let go. It's behind you."*

And yet, there it is—again. In the sudden sting of an old memory. In the familiar tightening of your chest when someone raises their voice. In the loop of thoughts that return every time you try to rest.

So let's ask it plainly: If the past is gone, why is it still alive inside you?

The answer, as always in UED, lies not in the story — but in the field.

Because the past isn't stored on a timeline. It doesn't sit in a folder labeled "history." It doesn't fade simply because the calendar changed.

The past persists because the ripple it created was never fully resolved.

You don't carry time — you carry unfinished alignment.

In Universal Energy Dynamics, every event in your life creates a ripple in the Structured Vacuum Energy (SVE) field. If that ripple returns to balance, it vanishes. It's complete. No emotion, no memory, no echo.

But if the ripple encounters resistance — emotional, psychological, energetic — it bounces back. It loops. It lingers.

That's what memory is. Not a record. A resonance.

So the past isn't behind you — it's around you, within you, still interacting with your present alignment. That's why certain places, people, or even smells can trigger it instantly — your ripple field recognizes the echo, and the loop reactivates.

You're not moving away from the past. You're carrying the unresolved ripple residue of it — until your field is ready to let it go.

This chapter is about seeing memory not as a film, but as a frequency. And understanding that real freedom doesn't come from running faster — but from realigning deeper.

Let's begin.

The Illusion of a Line

Time's Timeline vs Ripple's Residue

We're taught to visualize time as a straight line — one that moves forward, neat and unstoppable. Past on the left, future on the right, and your life like a dot sliding from one end to the other. It's a comforting picture. Logical. Sequential. Simple.

But reality is anything but.

The idea that the past is "behind" you is one of the deepest illusions we've inherited. It's baked into our language:

- "Leave it behind."
- "It's in the past."
- "Move on."

And yet, every one of us knows this is not how it works. You don't remember events like reading pages of a book. You feel them — as if they're still happening. A smell brings back the day your father left. A song revives a breakup like it's unfolding now. You step into a familiar room, and your chest tightens — you don't know why, but your body remembers something your mind has labeled "long gone."

Why?

Because in UED, time is not a flowing line. The past doesn't "pass." It resonates. It lingers.

Let's reframe.

Every experience you have — every conversation, heartbreak, joy, shock — generates ripples in your personal energy field. These ripples interact with the external field (Structured Vacuum Energy, or SVE). If the interaction completes — if there's no resistance, no unprocessed emotion, no mental friction — the ripple dissolves back into the field. Equilibrium is restored. You move on, cleanly.

But if the event is emotionally charged, misunderstood, resisted, or suppressed, the ripple doesn't vanish. It bounces. It reflects. It loops within your field. That's what we call "memory."

Not storage. Not footage. A living, vibrating echo.

That argument you had five years ago with someone you've never seen again? It's not back there in a file. It's here — still causing a little turbulence in your responses today.

Trauma is just an extreme version of this ripple imbalance. A traumatic event is not "over." It becomes a feedback loop that keeps replaying, not as thought, but as field disturbance. This is why trauma survivors often can't explain why they feel what they feel. They're not reacting to the present — they're echoing the unresolved ripple of the past.

Let's take a simple analogy. You drop a rock into a calm pond. It creates ripples. Over time, if left undisturbed, those ripples spread, weaken, and the surface settles again. But if you keep throwing more rocks, or the pond has boundaries that reflect waves back inward, the ripples stay. They crisscross. They amplify. And suddenly, the surface is in chaos — even though the original rock dropped long ago.

That's how most people live with their pasts. The rock dropped years ago. But the pond was never allowed to settle.

So when we ask, "Why can't I let go of this memory?" — the UED answer is:

"Because your field hasn't completed its return to alignment."

And if alignment never completes, the ripple doesn't stop.

Now here's the twist. Because the ripple still exists, it still influences your choices. That's how the past programs the present. Not because of fate, not because of destiny — but because your unresolved ripple is still shaping your field of perception.

This is why we repeat relationship patterns, fall into the same conflicts, attract the same type of people, make the same mistakes — even when we consciously *know better*. Our conscious choice is being filtered through a field shaped by unresolved echoes.

So let's return to the myth of the timeline.

There is no conveyor belt of time. There is no "behind" or "ahead." There is only your ripple field — disturbed or calm, aligned or misaligned.

And in that field, the so-called "past" is not back there. It's right here — in how you feel, how you react, and what you unconsciously align with.

The moment you realize this, your relationship with memory changes. You stop trying to "forget" the past. Instead, you start realigning your field — until that ripple no longer controls your now.

Memory, Identity, and Ripple Imprints

You Remember Because You're Still Aligned With It

We tend to think of memory as a mental snapshot — something stored in the "brain's hard drive" and pulled up like a file when needed. But memory, under Universal Energy Dynamics, is far more dynamic — and far more powerful.

Memory is not a dead archive. It's a live ripple signal still vibrating within your field.

When you remember something, especially something vivid or emotional, you are not simply "looking back." You're tuning in to an active ripple that has not yet collapsed. It's not behind you — it's aligned with you.

This is why a smell, a song, a texture can bring tears to your eyes. The ripple is already there in your field — and the sensory cue simply reactivates the loop.

Let's go deeper.

Every experience you've ever had leaves an energetic imprint. This imprint is not just psychological — it's vibrational. If the experience was emotionally neutral or if you processed it completely, that ripple dissolves. It becomes what we call a "resolved event."

But if the experience triggered fear, joy, shame, love, anger, or confusion — and especially if it was unexpressed or suppressed — then the ripple hangs around. It doesn't just "live in memory." It becomes part of your ongoing field signature.

This is how identity forms.

Identity is not who you are — it's the shape your ripple field has taken after years of repeated resonance.

That "shape" determines:

- The kind of people you attract
- The decisions you make automatically
- What triggers you and what soothes you
- Even what beliefs feel true to you

If you had a childhood where you were never heard, your ripple field may hold loops of neglect, invisibility, or over-efforting. As an adult, you may find yourself overexplaining, people-pleasing, or attracting relationships that ignore your needs — not because you choose it, but because your field is still aligned with the old ripple.

This isn't trauma in the dramatic sense. It's everyday misalignment — echoing in identity.

But trauma takes this ripple pattern to another level. Trauma is what happens when a ripple is so intense that your system freezes or shatters, and the alignment is never restored. It becomes a tight loop — replaying over and over. This is why trauma isn't remembered as a story — it's felt as a field disturbance. People say, "I didn't think it — I just panicked." That's ripple tension at work.

So what is healing?

Healing is not forgetting. It's not suppressing. It's not even talking endlessly. Healing is realignment.

To realign is to collapse the ripple — to return the field to a state where the echo no longer sustains itself. Once that happens, the memory may remain, but it no longer dominates your reactions. You remember — but you don't resonate.

Let's reframe this in UED terms:

- A **resolved memory** = a collapsed ripple loop. It no longer bounces around your field. No more feedback. No more ATT delay. It becomes neutral.
- A **stuck identity** = an unconscious resonance pattern. The same ripple keeps attracting similar responses from the world. It's like walking with a magnet stuck in your chest — drawing iron filings you don't want, but can't stop.
- A **wise person** = not someone who has read more books, but someone who has de-rippled most of their past. Their field is clean. Responses are fresh. There's no looping. They act from clarity, not compulsion.

This is why wisdom feels "light." Not because the wise have no past — but because they've collapsed its influence.

This also reframes the idea of "triggers." In the modern world, we say, "I was triggered," and often stop there — as if the trigger is the problem. But from a UED lens, a trigger is just a signal that a ripple loop still exists. You're not reacting to what's in front of you — you're echoing something inside you.

And so the path forward is not to avoid triggers, but to use them as field indicators. They show you where your alignment still needs work.

Let's pause with a metaphor.

Imagine you're playing a piano. Each key represents a past experience. Some keys are silent — fully resolved. Some produce music. And some, when touched, jolt you with dissonant, clashing sound — those are your ripple loops.

Most people try to avoid those keys. But true mastery comes when you can touch every key and produce harmony — because the dissonance has been transformed.

That's the gift of understanding memory through UED. You stop being a victim of your past — and become a composer of your present field.

In short:

- You don't remember because it happened.
- You remember because you're still aligned with it.

And when the alignment ends, so does the emotional pull. That's real closure — not forgetting the past, but freeing your present.

The Myth of "Letting Go"

Why the Past Only Leaves When You Realign With It We're told that "letting go" is a decision. That forgiveness is a choice. That moving on is a matter of willpower.

But the truth is, if it were that simple, no one would still be carrying pain years later.

How many times have you *told* yourself to move on? How many times have you *decided* to let go, only to find yourself angry again, grieving again, triggered again — just by a name, a voice, a smell?

It's not because you're weak. It's not because you lack control.

It's because "letting go," as society defines it, is a myth.

Let's demystify this.

From the UED perspective, you don't let go of anything — your field does. And that only happens when the ripple it's holding completes its feedback loop and collapses back into equilibrium.

In other words:

- You don't let go by force.
- You let go when the ripple no longer sustains itself in your field.

Let's go deeper.

When something painful happens — betrayal, loss, failure — it disturbs your ripple field. If that disturbance is processed, expressed, or understood meaningfully, the ripple weakens. It integrates. The field realigns.

But most of the time, we suppress. We deny. We distract. We try to "move on" intellectually while the ripple is still vibrating inside.

This is why time doesn't heal anything by itself. What it *can* do is lower your conscious engagement — but the ripple remains until alignment occurs.

Here's a simple example. Someone breaks your trust. Years pass. You haven't spoken to them. You believe it's over. But then you see them — or someone like them — and suddenly your stomach tightens, your heart races, your voice shakes. Where did that come from?

That's not memory. That's active ripple resonance. The field still remembers — because the realignment never happened.

Let's try another angle.

Imagine your ripple field as a pond. Something drops into it — an event. If you address the event clearly, the ripples fade. But if you resist, suppress, or deny the emotion, it's like placing walls in the pond. The ripples bounce off them. They don't dissipate. They loop.

Now society comes along and says, "Just let it go." But the pond can't be calm unless the structure changes — unless the walls are removed and the ripple is allowed to flow and dissolve.

That's what real letting go is: a structural field shift. Not a command.

This explains why:

- You can move on physically (new city, new job, new relationship), but emotionally still carry the same heaviness.
- You can convince your mind that "you're over it," but your body reacts instinctively when that chord is struck again.
- A place, a smell, a song can reawaken a long-dead feeling — not because the past returned, but because your field was never freed from its echo.

Now here's the empowering twist: You don't need to suppress the ripple. You need to feel it through — until the field completes its alignment and the energy naturally resolves.

Forgiveness, in this view, is not about "agreeing" with what happened. It's not even about the other person. It's about freeing your field from the loop they left behind.

Letting go is not mental. It's energetic.

And it's only possible when:

- You acknowledge the ripple
- You stop feeding it new resistance
- You allow the feedback loop to resolve through awareness, truth,

and release

In Universal Energy Dynamics, the core principle is this:

"The past doesn't haunt you. Your ripple still responds to its unfinished feedback."

Once the feedback stops — once the ripple no longer echoes — you don't have to let go. It simply isn't there anymore.

This is why some people feel like they've *suddenly* healed after years of carrying something. A moment of insight, a shift in awareness, a safe conversation — and the field releases. Not because they "tried" harder, but because alignment finally happened.

It's not about control. It's about coherence.

So the next time someone tells you to let go, you might respond:

"I'm not holding on. My field is still learning how to stop echoing."

That's not weakness — it's profound awareness.

And ironically, the moment you stop trying to force letting go — and instead begin to observe, align, and allow — is the moment it begins to let go of you.

Closing Reflection:

You're Not Haunted by Your Past — You're Still Echoing It

We've been taught to treat the past like something linear. A place we leave behind. A sequence we outgrow. A ghost we outrun.

But you're not running from the past. You're vibrating with it.

You're not haunted by your past — you're still echoing it.

The painful memory that resurfaces without invitation... The automatic reaction that doesn't match the present moment... The story you keep repeating, even when you say you've moved on...

These are not evidence of weakness. They are the signatures of ripple residue — unresolved loops still influencing your field.

You cannot outrun the past, because in UED, there is no running — only realigning. Your field holds the echo until the feedback completes. Your identity loops the story until the vibration dissolves.

This is why time doesn't heal all wounds — alignment does. You don't "let go" of the past through forgetting or distraction. You dissolve it by changing your resonance — by realigning your field until that ripple no longer finds a match inside you.

And when it no longer finds a match...

- The trigger fades.
- The memory softens.
- The energy dissolves.

This is the true nature of closure: Not forgetting the past, but becoming so deeply aligned in the present that the past has no foothold left.

Because the past was never a road behind you — It was always a field within you, influencing and being influenced by every step you take. And now that we've reframed memory...Here's a question for what's ahead:

If the past still lives inside your field, then what about the future? Could it already be forming within you — right now — as a ripple waiting for alignment?

In the next chapter, we'll explore the most misunderstood dimension of all: The future is not ahead of you. It's already in motion — inside your intent.

Let's go meet it.

"The past is not what you remember. It's what your ripple hasn't finished responding to."

Chapter 5: Future Doesn't Exist Yet

"What if the future isn't waiting for you — it's waiting on you?"

We've been raised to believe the future is a place — a dimension ahead of us, unfolding like pages in a cosmic calendar. We prepare for it. Predict it. Fear it. Sometimes we chase it. Sometimes we surrender to the idea that "whatever's meant to happen will happen."

But pause for a moment. What if that entire premise is flawed?

What if the future isn't something "out there" — not a destination ahead on a timeline, not a scene waiting behind the curtain of time?

What if the future is not waiting for you... It's waiting on you.

In Universal Energy Dynamics, the future isn't real — not in the way we conventionally understand it. There is no forward motion in the universe. There is no ticking clock pulling you toward some predetermined point. There's only a still field — a silent canvas of ripple potential — and your alignment with it.

That alignment — or the lack of it — is what decides what you experience next. Not destiny. Not probability. Not time. Just pure ripple resonance.

The so-called future isn't a certainty unfolding. It's a possibility attempting to align.

Like a tuning fork trying to find its echo in your field. If it finds resonance, it becomes real. If not, it stays unrealized.

This reframes everything:

You don't walk into your future — you magnetize it.

Your next moment isn't guaranteed by time — it's granted by alignment.

The more coherent your intent, the clearer the future's shape. The more scattered your field, the more uncertain your path.

So the real question isn't "what's going to happen?" It's: "What ripple am I radiating — and what is it attracting back?"

Let go of the timeline. Step into ripple-time. Because in truth, the future doesn't exist — until you do something that brings it into being.

The Illusion of a Waiting Future

Why "What's Next?" is the Wrong Question

Ask any student, professional, or parent, and you'll hear it: "What's next?" What's next after school? After this job? After marriage? After retirement?

We are conditioned from early childhood to forecast life — as though it were a weather system just a few days ahead. Society rewards long-term planners. Education systems demand career roadmaps. Corporations rely on five-year strategies. Even spirituality now comes with a vision board.

The assumption beneath all of this? That the future exists ahead of us, like a road we're destined to walk — and our job is to reach it.

But what if that's not how reality works?

The Stillness Beneath Movement

Universal Energy Dynamics (UED) flips this paradigm entirely. In UED, nothing in the universe moves. What we call "motion" is simply changing ripple alignments in a still Structured Vacuum Energy (SVE) field.

So if nothing moves... how can the future be "ahead"?

The idea of a future "coming toward you" or you "moving into it" collapses — because there's no universal conveyor belt pushing you along a timeline. There is only your field of intent, and how it aligns — or misaligns — with the ripple dynamics of your surroundings.

The future, then, isn't an *event ahead.* It's a field possibility that activates *only when your ripple pattern supports it.*

There Is No Road — Only Resonance

Imagine a violinist standing in a quiet hall. Around her are thousands of instruments — guitars, drums, pianos. But only one violin string starts to vibrate when she plays. Not because it was destined. Not because time had arrived. But because resonance aligned.

This is the future in UED: not a road you travel, but a resonant point that activates in your field. Until your ripple matches it, it simply does not exist.

The future doesn't wait for you. It becomes real *only if you're aligned enough to meet it.*

Planning vs Attuning

Now, this doesn't mean you stop planning or working toward goals. It means you stop expecting the future to obey the calendar.

Conventional thinking says:

- "If I put in X effort, I'll reach Y outcome by Z date."

But UED asks:

- "Is your field attuned to receive that outcome at all?"

This is why some people try for years and see no result — while others seem to "manifest" things effortlessly. It's not luck. It's not effort alone. It's ripple alignment.

And this also explains why some goals fall apart despite everything going right. Because the ripple of your effort never found its matching wave in the field. The road didn't reject you — it simply never existed.

Time Doesn't Pull — Alignment Does

In most people's minds, time is a flow:

- Today pushes into tomorrow
- Tomorrow pushes into next week
- And so on.

But in UED, there is no push. There's only pull — and it's based on alignment.

A future event doesn't "come" because time moved forward. It appears because your ripple resonance became suitable enough to make it real.

So instead of asking "What's next?" Ask:

- What am I attuned to?
- What is my field radiating?
- What ripple pattern am I sustaining?

Because *that* — not a date on the calendar — is what determines your next experience.

The Wave That Never Arrives

Think of a surfer staring at the sea, hoping a perfect wave will come. The ocean isn't timing itself to deliver anything. It's pulsing with possibilities. The surfer can only catch the wave when ready — with balance, presence, and alignment.

The wave isn't "next." The wave is — and it's only yours if you match it.

So too with the future.

In short:

- The future doesn't exist "out there."
- There is no timeline unrolling.
- There is only ripple readiness.

And the great illusion of modern life is that we're marching into a future we assume is waiting. But reality doesn't wait — it responds.

So the better question is not: "What's next?" But: "What am I aligned with — right now — that's ready to emerge?"

Possibility vs Probability vs Ripple Alignment

What Determines What Comes "Next"

In the conventional worldview, everything about the future is cast in terms of possibility or probability. Possibility asks, *"Can this happen?"* Probability asks, *"How likely is it to happen?"*

From career moves to weather forecasts to economic predictions, we are taught to measure the future through statistical lenses. If something is *possible*, we plan for it. If something is *probable*, we expect it. That's how modern civilization runs — on extrapolated likelihoods.

But there's a hidden assumption in all of this: That time will unfold smoothly and outcomes will arrive linearly if we just calculate well enough.

UED doesn't oppose this model — it simply completes it.

Because possibility and probability don't determine reality. Alignment does.

The Classic Two: Possibility and Probability

Let's clarify the first two lenses:

- **Possibility** is the **open canvas** — anything that can theoretically happen. Example: It's *possible* you could start a company, win a Nobel Prize, or move to Mars. Possibility is vast — it's imagination unbounded by constraints.
- **Probability** is **likelihood** — what's *most* expected based on current trends, actions, and available data. Example: If you've been writing daily, it's *probable* you'll finish your book soon. If the clouds are dark, rain is likely. Probability is reason's attempt to guess the next

ripple based on past patterns.

But life often breaks both these frames.

Sometimes the most *probable* events never happen. Sometimes the *least likely* ones hit you like a lightning bolt.

Why? Because neither possibility nor probability has control over field alignment.

The UED Third Lens: Ripple Alignment

In UED, everything that emerges in your experience is not because it was possible or probable — It's because it aligned.

Your ripple field — a dynamic structure shaped by your thoughts, emotions, actions, memories, and intent — is constantly interacting with the ambient ripple environment of the universe (SVE).

What manifests next isn't what's most likely — It's what finds resonance.

This is why:

- You may do everything right for years, yet nothing "clicks."
- You might get a breakthrough opportunity *unexpectedly* — from someone or somewhere unrelated.
- You meet a person at "just the right time," even though the odds were minuscule.

These aren't miracles or coincidences. They're ripple alignments — fields meeting at the right energetic pitch, irrespective of logic.

Why High Probability Sometimes Fails

Think of a talented student applying to a dream college. They have the grades, the portfolio, the recommendation letters. All statistical models say: "Highly probable." But the admission doesn't happen.

From the outside, this looks like injustice or randomness.

From UED: the student's ripple intent field may not have aligned with the field of that institution at that point in time. The inner state — fear, expectation, ego, or desperation — may have disrupted the coherence.

Misalignment = no materialization.

Just as a song only plays when the needle finds the groove, the outcome only emerges when resonance is clean.

Why the Improbable Sometimes Happens

Now flip the script.

A beginner musician uploads a raw video online — no PR, no fans, no big expectations. It's unlikely anyone sees it. But it goes viral.

Why?

Not because of luck. Because somewhere, the field was ripe — someone's ripple resonance matched this unknown voice, and the alignment caused a cascade of ripple reactions.

That video wasn't likely. But it was resonant.

In UED, resonance doesn't ask permission from statistics. It activates potential when conditions in the field are met.

The Ripple Precursor Principle

This leads us to a profound insight in UED:

Every future outcome is seeded by a resonance in your present intent.

You are not predicting or pulling the future. You are vibrating — and the universe is matching.

The seed of tomorrow is not what you imagine, or even what you plan. It is what you are currently aligning with — emotionally, mentally, energetically.

This is why many spiritual traditions emphasize "being" over "doing." Intent, in UED, is not just hope — it's ripple structure formation. It is the geometry of what's coming, whether you see it or not.

Previewing ATT: Alignment Threshold Time

But here's the twist — just because alignment has started doesn't mean the result will appear instantly.

There is always a delay — the time it takes for your ripple pattern to fully synchronize with the environmental field.

This is called ATT — *Alignment Threshold Time.*

You may be aligned in intention today — but the universal field requires ripple maturity. Like a tree from a seed, coherence must reach a threshold before manifestation.

ATT is what people mistakenly call "timing" or "destiny." It's neither fate nor delay — it's field latency.

Just because it's aligned doesn't mean it's due. But without alignment, it's never due — no matter how long you wait.

Final Thought

Don't chase possibility. Don't get obsessed with probability. Focus on your ripple alignment — the only real determinant of what comes next.

Because the truth is:

The future isn't something to forecast — it's something you tune into.

Manifestation, Intent, and ATT

Why Desire Isn't Enough — and Action Isn't Always Control

Manifestation. A word that has been glorified, mystified, and misused more than almost any other in the world of self-help and spirituality.

The popular idea:

"If you want it, think about it. Feel it. Believe it. And the universe will deliver."

The problem? People think, feel, believe... and yet — nothing happens. Or worse, the opposite happens. And soon enough, manifestation is either dismissed as pseudoscience or blamed on the individual for not being "positive enough."

UED enters this conversation with both humility and clarity.

It doesn't mock intention — it honors it. But it also doesn't romanticize desire. Because desire without ripple alignment is just noise in the field.

Misconceptions of Manifestation

Let's break it down.

Mainstream manifestation philosophy teaches that thought = creation. But in UED terms, thought is just the first ripple, not the outcome.

A desire you form — no matter how strong — is merely an internal vibration. For that vibration to become real, it must enter into resonance with the external field.

That's not instant. That's not magic. That's physics of alignment.

Just like you can tune a guitar string, but until the air around it and the instruments nearby catch the tone, the sound doesn't propagate as music. Similarly, your intent must not only be clear — it must be field-compatible with what exists outside you.

Introducing ATT: Alignment Threshold Time

This is where most people get trapped.

They think: "I'm clear in my vision. I've taken action. Why hasn't it shown up?"

Because they misunderstand the lag — the *Alignment Threshold Time (ATT).*

ATT is the time it takes for your internal ripple to synchronize with the universal field enough to trigger a consequence.

It's not delay. It's not punishment. It's not resistance. It's natural latency — like waiting for water to boil or fruit to ripen.

Every ripple needs maturation time in the field. Every outcome needs to build enough tension or coherence before it clicks into the visible world.

This is why:

- Sudden success is never really sudden — the ATT was just silent.
- Manifestation doesn't fail — it just hasn't crossed the threshold.
- The universe is not withholding — it's waiting for you to match up.

Desire Without ATT Awareness = Frustration

A desire, no matter how burning, doesn't override ATT. Trying to manifest faster than alignment allows is like trying to microwave a seed into a tree.

You get heat. Not growth.

This is why many people give up halfway. They misinterpret the quiet period as failure — not realizing it's the exact midpoint of field transformation.

And this is where inner disturbance begins:

- Doubt
- Impatience
- Self-sabotage
- Over-efforting

All of which disrupt your ripple field — thus extending your ATT even more.

Premature Action = Ripple Interference

Another mistake? Acting too soon.

Just because you've *decided* doesn't mean your field is ready. Taking action without ATT maturity is like plucking a fruit before it's ripe.

The consequences:

- Projects collapse
- Relationships don't hold
- Deals fall through
- The effort feels exhausting instead of exciting

Why? Because your ripple wasn't ready, and neither was the field.

You forced what needed to form.

This is not passivity — it's precision. Knowing when to act is not based on time — it's based on ripple *readiness*.

Analogy: Sowing Seeds and Holding Fields

Imagine a farmer sowing seeds. Does he sit and stare at the soil every day, waiting anxiously? No. He tends to the field — ensures water, sunlight, protection — but trusts the process.

Why?

Because he knows growth has its own ATT. He cannot pull the sprout out of the soil by sheer desire.

Similarly, in UED:

- The moment you form a pure intent, you plant a ripple.
- But the field in which it grows must remain stable and coherent.
- Any distortion (doubt, fear, anxiety, pushing) dampens the ripple.

So your job is not to control what happens — Your job is to sustain alignment until it's inevitable.

Ancient Echo: Let Go, Yet Align

This brings us full circle to ancient wisdom that often sounds contradictory:

"Let go." "Detach." "Do the action, not the fruit."

People mistake this as disinterest or inaction. But through the UED lens, it becomes crystal clear:

Letting go doesn't mean giving up the desire. It means releasing the friction while maintaining alignment.

- Detachment isn't apathy.
- It's ripple *cleanliness* — allowing resonance to form without emotional static.

This is how monks manifest serenity. This is how visionaries hold dreams across decades. Not by pushing — but by holding the field without leakage.

Reframing Effort: Maintain Ripple Coherence

So what *should* you do?

Effort is not irrelevant. It just needs redirection.

- Don't obsess over the outcome. → Refocus on maintaining your ripple purity.
- Don't push the world. → Refine the quality of your field response.
- Don't act to force change. → Act to express coherence — that's what triggers response from the field.

In essence:

Stop trying to create the future. Start holding the pattern of the future — right now.

Because everything you're waiting for is already in formation. And your only real job is to not disturb it.

Time is Not Pulling You Forward — You're Pulling Forward Outcomes

From Timeline to Rippleline

We've all heard it:

"Time will heal." "Time will tell." "With time, things will happen."

These phrases are baked into our cultural mindset, subtly suggesting that time itself is an agent — a force moving us forward like a train on invisible tracks, dragging our experiences along.

But what if the train isn't moving?

What if *you* are the one pulling experiences into reality — not by walking through time, but by realigning your field?

The Timeline Illusion

Mainstream understanding of time paints it as a linear track, stretching from past to future, with you as a passenger riding from station to station. Today becomes tomorrow, tomorrow becomes next week — and the train moves forward no matter what you do.

But under Universal Energy Dynamics (UED), that's a complete illusion.

There is no train. There is no track. There is no forward motion.

There is only a still field — and a constantly shifting ripple pattern within it. Your intent. Your thoughts. Your desires. They're not *on* a timeline — they're realignments inside a ripple field.

The future isn't pulling you toward it. You're pulling forward whatever field configuration you're aligned to receive.

Future Is Not a Date — It's a Mirror

Imagine you're standing in front of a mirror — but this mirror only shows a reflection when your internal ripple becomes clear enough.

The future isn't a date on the calendar. It's a field echo — waiting for the right resonance to appear.

You don't move toward it. You *tune into it.*

Until then, it's not "waiting" — it simply doesn't exist in your experience yet.

The future doesn't arrive because time passed. It arrives because alignment occurred.

The Rippleline, Not the Timeline

So let's toss out the metaphor of the timeline altogether. Replace it with what UED proposes: the Rippleline.

Here's how it works:

- You're not moving across time.
- Your field configuration changes, and this shift brings new realities into phase with your consciousness.
- You don't "travel" to the future — you become aligned with it, and therefore it becomes visible and experiential.

From this view:

- Growth isn't passage through time — it's field reorganization.
- Aging isn't time dragging you — it's ripple fatigue and entropy accumulating in misaligned zones.
- Success isn't reaching a point in time — it's activating a coherent possibility through intentional resonance.

A Rearranging Puzzle, Not a Track

Picture this: You're standing on a huge floor made of shifting tiles. Each tile is a possibility. As your ripple pattern changes, the tiles rearrange themselves. What appears in front of you isn't where you were going — it's what your field unlocked.

In this analogy:

- There's no future "out there."
- There's just what your field can currently handle.
- Every shift in belief, emotion, awareness — nudges the tiles, reconfiguring the field.

The "future" then is simply the next tile that fits your new shape.

The Paradox: Stillness Creates Change

Here's the strange but beautiful paradox of UED:

You're not moving — your ripple state is. But because your perception is bound to your ripple state, it *feels* like movement through time.

That's why:

- A long conversation can feel like a minute.
- Waiting five minutes in pain can feel like an hour.
- A flash of inspiration can reorder your entire life.

Time didn't change. Your ripple tension did — and with it, your felt experience of the now.

This also explains why:

- Stillness in meditation expands your sense of time.
- Flow states collapse the sense of waiting.
- Grief distorts hours into eternities.

Time isn't flowing. Your resistance to reality is fluctuating.

Coherence is the Catalyst

Let's reframe the "arrival" of the future one last time.

What you call the "future" only shows up when:

- Your internal ripple becomes *coherent enough*
- Your emotional and cognitive field becomes *quiet enough*
- Your resistance and craving both reduce enough to allow a new pattern to emerge

This is why spiritual traditions urge patience, presence, and surrender — not because they lack ambition, but because they understand manifestation is not forced — it's invited.

You don't rush the future. You receive it when you are finally a fit for what's meant to arrive.

And this — not time — is what changes lives.

TL;DR Recap:

- Time doesn't move. You don't travel through it.
- You realign internally — and outcomes *appear* externally.
- The future is not a road ahead — it's a pattern awaiting a match.
- The only "progress" is ripple purification and field coherence.

Time is not pulling you forward. You are pulling forward whatever you're tuned to experience.

Closing Reflection

"You don't walk into your future — you become it."

Let's pause on that for a moment.

Because that single sentence challenges everything we've been taught. Not just in school, but in how we live, plan, and hope.

The truth is: there is no future waiting for you somewhere up ahead. There is only your field. Your resonance. Your ripple pattern.

Nothing "next" is fated to arrive. Nothing unfolds just because the clock keeps ticking. Instead, your internal state pulls forward whatever can match it.

In Universal Energy Dynamics, the concept of a timeline is replaced by a rippleline — not a path you follow, but a pattern that unfolds because of you. And that's not mysticism. It's mechanics. The future is simply a field response to your current alignment.

So when people ask, *"What's coming next?"* The better question is: "What am I currently aligned with enough to allow into experience?"

That's what determines emergence — not probability, not prediction, not even preparation in the conventional sense. Alignment is the gatekeeper.

Just like the past doesn't leave you until its ripple is resolved, the future doesn't arrive until your field is ready.

And when it's not ready, what do you feel?

- Anxiety: from premature expectations misaligned with your current field.
- Overthinking: from trying to manipulate timing instead of refining coherence.
- Frustration: from action without ripple readiness.

So what's the way forward?

Not to chase the future. But to fine-tune your present field so clearly, that only the most coherent, precise, and resonant outcomes are *pulled forward* — naturally.

That's the paradox of progress in UED: You don't move into the future. You dissolve the misalignments that are preventing it from emerging.

And in doing so, you don't walk into your future.

You become it.

Next up: If time isn't moving, and both past and future are just ripple residues inside you...Then what exactly is the Now? Is it a pause? A blur? A portal?

Possibility vs Probability vs Alignment

Aspect	Possibility	Probability	Alignment (UED Lens)
Definition	Anything that *can* happen	What is *likely* to happen	What *will* manifest due to field resonance
Basis	Imagination or potential	Past data, trends, statistics	Ripple coherence and present intent
Scope	Infinite — unbounded	Filtered — based on assumptions	Dynamic — shaped by real-time ripple configuration
Role of Mind	Dreamer	Analyst	Aligner and tuner of personal field
Example	"It's possible I become a filmmaker."	"Based on trends, it's unlikely."	"If my field aligns, it can happen — regardless."
Emotional Effect	Hope, fantasy	Doubt, planning, risk analysis	Clarity, surrender, confidence through resonance
In UED Terms	A ripple that *could* initiate	A ripple that *might* echo	A ripple that *is already syncing* with outcomes

Ripple Precursor to Manifestation Map

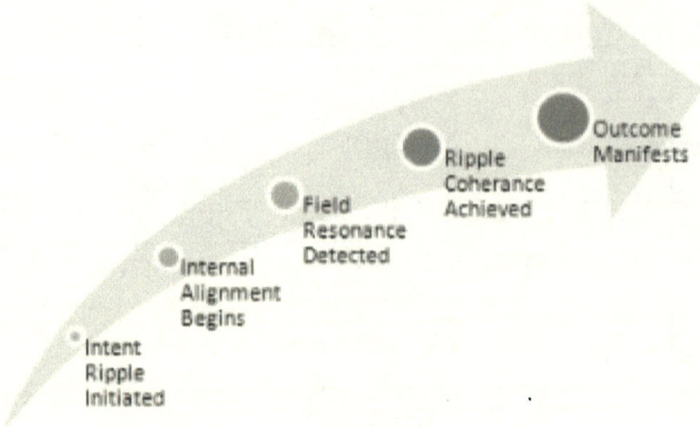

ATT Curve — Why Some Desires Take Time

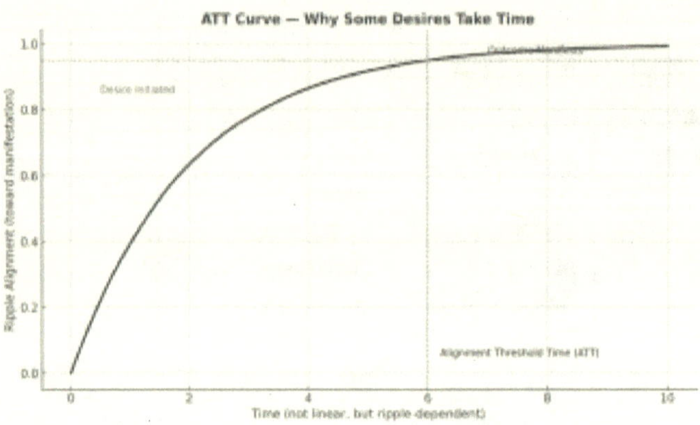

Chapter 6: "The Present Isn't a Point"

The Now is Not a Dot — It's an Interval of Interaction

"If the present were just a point, how could anything happen in it?"

People talk about the present as if it were a dot on a ruler — clean, singular, razor-thin. They say "live in the now" like it's a pinpoint you can plant yourself on, perfectly balanced between the past and future. But let's get honest for a moment. Can anything truly occur in something that small?

If the now were a point, there'd be no room for choice. No space for perception. No moment to even feel that you are alive. And yet, everything that has ever happened to you has only happened in what we call the present. That's not a point. That's a field.

Under UED, the present isn't a pause or a blink. It's not where the timeline stops. It's a zone of interaction — the live space where internal ripples (your awareness, intent, emotion) meet external fields (events, environments, energies). This is the interval of consequence — where realignment occurs and outcomes begin.

Think of it like a window opening to the only place you can actually engage reality. Not yesterday's memory. Not tomorrow's hope. Just this... ripple field, buzzing with feedback, tension, potential.

In physics, the smallest definable instant is called the "Planck time." But even that's not where your life is. You don't live in Planck units. You live in response units — ripples that carry the consequences of your choices and reflections of your reality.

So let's move beyond mindfulness memes and time management clichés. This chapter is not about "being present." It's about understanding that the present is an energetic phenomenon — not a chronological tick. And the better you sense, stretch, and stabilize your field in this interval, the more fully you'll live.

Because the now isn't a dot. It's the only door that's ever truly open.

The Misunderstood Now

Why "Living in the Moment" Is Confusing

"Live in the now," they say. It's one of the most overused prescriptions in the world of self-help, wellness, and pop-spirituality. But what does that actually mean? For some, it sounds like an invitation to hyperfocus — to dial into the current second with such intensity that nothing else exists. For others, it's just a feel-good slogan, like putting a sticky note on your fridge that says *"Breathe."*

The problem isn't with the advice — it's with the underlying assumption. We've been taught to imagine the "present" as a mathematical dot, a precise zero-width point between the past and the future. It sounds elegant. It sounds tidy. But it's also completely misleading.

Because if the present were really just a dot, nothing could actually happen in it. Action takes space. Response needs a window. Reflection takes more than zero time. Even a thought or a heartbeat spans a range. So how can a moment of such infinitesimal size contain the entirety of your experience?

The truth is, it can't.

Under mainstream physics, we inherited a timeline view of reality — the idea that life is a continuous film reel, with frames moving from past to present to future. The present, in this model, is the frame that's currently in the projector. It appears and vanishes so fast, you can barely notice it.

But under UED (Universal Energy Dynamics), this image breaks down. Because UED doesn't treat reality as a flowing river of time. It treats it as a still field of structured vacuum energy — where nothing moves, and all "motion" is ripple interaction. In this view, the present is not a frame or a point. It's a wavefront — a field of live interaction between internal intent and external ripple reality.

The "now" is where alignment happens. And alignment, by its very nature, takes interval — not instant.

Let's use an analogy.

Imagine you're tuning an old radio. You're slowly adjusting the dial to find a station. As you move the dial, you pass through static, interference, and finally hit the sweet spot — clear signal. That clear reception isn't a dot. It's a bandwidth — a frequency range in which the internal circuitry of your radio aligns with the broadcast waves in the environment. That's what the present is under UED: the coherence band.

Now, think about when you're truly "present" — in deep meditation, in the zone during a performance, in a meaningful conversation. Does it feel like a blip? Or does it feel like a field? A stretch of space and time where you feel more real, more connected, more alive?

That's because your ripple field is actively engaged — and fully interactive — within a coherent zone of the universal field. The "now" you feel is not an instant. It's an interval of perception and response.

But most people misinterpret "living in the moment" as either:

- Impulsive reaction ("Just do it!"), or
- Mental stillness ("Don't think — just be here").

Neither of these captures the real nature of the present. Because being present is not about switching off thought or riding a dopamine high — it's about ripple awareness. It's about recognizing that the quality of your experience depends on how well your internal field is interacting with the outer field right now.

That's why some people can feel fully alive in a few seconds, while others can go hours feeling numb. The clock doesn't determine presence — alignment does.

Let's take it one step deeper.

When you stub your toe or hear sudden news, you feel the present *intensify*. Time "slows down," but your awareness expands. Your field rushes to align with the ripple shockwave. That's presence, but not peace.

In contrast, when you're in love, in flow, or deep meditation, time slows for a different reason — because your field stabilizes. The present widens not because of urgency, but because of coherence.

In both cases, you are not *in* a point. You are *inside a field of resonance.*

This is the misunderstood now: the idea that it is small, fleeting, and passive. But in reality, it is large, consequential, and alive. The present is not a pause. It is the pulse of perception. The only pulse that matters.

And once we shift from thinking of now as a tick on a clock to a ripple window of consequence, our understanding of life — and time — fundamentally changes.

The Present as a Field

Now = Ripple Response Zone

If the past is a residue echo and the future is potential not yet aligned, then what is the only place where life actually happens?

The present.

But not the present you learned in school — not the frozen frame between yesterday and tomorrow. Not the fraction-of-a-second tick that your smartwatch records. Not the hollow "now" that floats in quotes on Instagram.

Under UED, the present is not a point — it's a field.

Let's call it what it is: the Ripple Response Zone. The one interval in which reality is not remembered or imagined — but engaged.

In Universal Energy Dynamics, the universe is not moving through time. Instead, it is a structured, stable background — a field of still but responsive energy (SVE). Motion, change, and causality aren't happening because of time. They happen because of ripple interactions — your intent, your perception, your action — encountering the external ripple field.

This means:

- The past is the ripple residue of earlier interactions still echoing in your field.
- The future is the ripple potential, shaped by alignment readiness.
- And the now is the active interface — the bandwidth where response becomes reality.

In simpler terms: you live only in this field. Everything else is replay or prediction.

When you feel something, decide something, or experience something — it's not occurring in "time" per se. It's occurring in this dynamic interval of real-time alignment between your internal ripple field and the surrounding environment. You're not passing through it — you are actively shaping it.

Why All Sensation Is Field-Based

Think of pain. You don't feel it from the past — you feel it in the moment it is registered, even if caused earlier.

Think of memory. You don't feel nostalgia from the past — you feel it now, through the resonance of that ripple still active in your field.

Think of emotion. Anger, love, awe — these aren't reactions to what *was*, or what *might be*. They're field signals — born, registered, and resolved in the ripple response zone of now.

Even your thoughts, despite dancing with past data or future plans, happen only as present field fluctuations — interactions within your active consciousness band.

So it's not just that the present is where you are — it's the only place you can actually be.

Why High Awareness Expands the Present

Have you ever noticed how moments of crisis seem to slow down?

When something sudden happens — a car skids, someone faints, or a child runs into the street — time seems to stretch. You're aware of every detail: your breath, the sounds, the ground beneath your feet. The world doesn't freeze. You just enter full presence.

That's not because the clock slowed down. It's because your ripple field widened its alignment zone. In that state, your awareness becomes a high-resolution sensor, tuning deeply into the ripple dynamics around you. You're not reacting late. You're interacting precisely.

In that moment, "now" is not a second — it's a symphony of responses, happening across your cognitive, emotional, and physical fields.

This explains why elite athletes, surgeons, or performers in flow state don't feel time. They're not in a timeline — they're in a fieldline — a present so fully engaged that the experience of duration vanishes. In UED terms: the ripple alignment bandwidth becomes highly coherent, and resistance drops. That's when the field feels timeless.

Conversely, when you're bored, anxious, or disconnected, time feels slow and heavy. That's because your ripple field is misaligned — and the resistance creates friction. You're not interacting — you're dragging.

The Now Is a Realignment Engine

You don't *just exist* in the present. You are constantly realigning in it.

Every breath adjusts your biological rhythm. Every thought tunes your perception filter. Every feeling redirects your emotional coherence. Every pause, every decision — is a ripple reset.

This is the power of the present: not that it's available, but that it's *interactive.* The past may haunt you, and the future may tempt you, but only the present can heal, reshape, and redirect you.

That's why in UED, "living in the present" isn't about silencing thought or being mindful in the Instagrammable sense. It's about recognizing that the now is your operational zone — the dashboard where your inner field meets the universe, constantly adjusting based on how well you align.

When you say something wise, forgive someone, feel a surge of joy, take a deep breath, or decide not to react — it all happens in this ripple response zone. And what emerges from it becomes your next echo.

So, no — the now is not a pin on a map. It's the landscape you're standing on.

And in that landscape, you're not a passive traveler. You're the ripple tuner — sensing, shifting, and shaping the only space that's ever real: this one.

Expanding or Shrinking Now

Why Time Feels Fast or Slow

"If you say 'time flew' or 'time froze' — what actually changed?"

We grow up thinking time flows at a constant rate. Sixty seconds is always sixty seconds. But internally, experientially, that simply isn't true. A five-minute wait in a hospital feels like an hour. A three-hour dinner with someone you love vanishes in a blink. What changed?

Not the clock.

What changed was your ripple field — and the width of your Now.

ATT and Awareness: What Shapes the Width of Now

In the UED model, ATT (Alignment Threshold Time) is the internal delay it takes for your ripple field to sync with an external field. It determines how deeply you can experience an interaction. When your field is stable and open, ATT lowers — and your now "widens." You sense more, feel more, respond with presence. When your field is tense or disrupted, ATT increases — and the now collapses. Your awareness gets choked by interference.

This is why time feels "fast" when you're anxious, and "slow" when you're centered. Time is not flowing. Your now is stretching or shrinking.

When Trauma Compresses Now

Consider the mind during trauma. Everything becomes hyper-reactive. You lose a sense of continuity. Decisions become impulsive or frozen. People say, "I blacked out," or "It all happened so fast." This isn't memory loss — it's a narrowed now field.

The ripple field is so disturbed, so overrun with feedback loops, that your experience collapses into spikes of sensation and panic. Time appears to vanish, but it's really your perception of sequence that breaks down.

In UED, this is ripple overload. The system isn't designed to align in that state — it's too turbulent.

When Flow Expands Now

Now shift the lens to a creative flow state, or deep meditation. You're not watching the clock. You're not worried about outcomes. You're fully immersed in the ripple field of the task or thought or presence. People call this the "zone."

In this state, your ripple field is coherent. ATT is minimal. Feedback and intent meet without delay. This gives rise to what feels like timeless experience — not because time stopped, but because there was no resistance to awareness.

You were the ripple, aligned.

When Love Makes Time Eternal

Love — real, conscious connection — often expands now more than anything else. In the presence of someone who resonates with your field, alignment becomes natural. There is less need to process and more freedom to be. That deep exhale? That ease of sharing without fear? That's your ripple field breathing.

Time, here, is not even a concern. Not because you're ignoring it — but because you're not creating delay. You're not resisting or avoiding. You're simply in coherence.

The Ripple Width of Now

Let's place this in a tangible framework. The following comparison shows how your state of being alters the width of now, the quality of ripple alignment, and your felt sense of time:

State of Experience	Now Field Width	Ripple Alignment	Time Perception
Trauma	Narrow	Disrupted	Frozen / Too Fast
Anxiety	Fluctuating / Narrow	Distorted / Unstable	Erratic / Rushing
Normal Daily Awareness	Moderate	Average / Functional	Clock-bound
Focused Work / Productivity	Widening	Increasing Coherence	Efficient / Slower
Meditation / Flow State	Wide / Stable	High Coherence / Minimal Resistance	Timeless / Extended
Love / Deep Connection	Wide / Expansive	Full Resonance / Low Resistance	Present Feels Eternal

This table isn't just psychological. It's energetic. It maps directly to how much of the Now Zone your mind is able to access based on your internal ripple alignment.

Analogy: Widening the Window of Now

Think of the now not as a moment you're walking through, but a window of detection. The wider that window, the more information, sensation, and clarity passes through. When your field is coherent, that window widens — and time feels richer. When your field is overwhelmed, that window narrows — and time either rushes by unnoticed or explodes into panic.

You're not running out of time.

You're just not fully inhabiting it.

Closing Insight

Your goal isn't to "slow time" — it's to widen your field so that time stops being a pressure and starts becoming a platform. When the now is wide, you act with grace. When it's narrow, you react with fear.

So the real skill is not time management — but field alignment.

Ready to explore what "Now" is truly made of?

Decision-Making, Presence, and Ripple Clarity

Your Best Choices Happen in a Wide Now

People often say, "Trust your instinct," or "Be present when deciding." But what does that actually mean?

Let's step back from the philosophical and zoom into performance, leadership, and creativity. Whether it's a tennis champion in a tie-break, a crisis leader making a call under pressure, or a jazz musician improvising live — their magic comes not from prediction, but from presence.

Not just any presence — but a *wide*, clear now.

The Performance Paradox: Presence Without Overthinking

At peak performance, the mind is not chattering. It's not anxiously weighing past failures or hypothetical futures. It's *tuned*. It's absorbing inputs and releasing responses in real time, like a perfectly balanced wave.

This is ripple coherence in action.

In UED terms, what happens here is:

- The internal ripple field is aligned — no unresolved turbulence from the past
- The external field is being read accurately — minimal interference
- The ATT (Alignment Threshold Time) is low — decisions arise effortlessly

Great decisions are never really *calculated* — they are *perceived*. The clarity required doesn't come from thinking harder, but from *sensing without noise*.

This is why elite performers often describe the moment as "still," or "slowed down." Time didn't change. Their now widened.

Real Presence Is Not a Gimmick — It's Ripple Hygiene

"Be present" has become a commercial cliché. There are mindfulness apps, timers, ambient playlists, meditation cushions — all trying to manufacture presence.

But real presence has nothing to do with gadgets or practices. It's a field state.

If your ripple field is:

- Jammed with unresolved echoes (past trauma)
- Scrambled by anxious projections (future overload)
- Distracted by external noise (overstimulated environment)

...then no amount of "breathe in, breathe out" will make you truly present.

Presence, in UED, is not a technique — it's a consequence of internal clarity.

A clear field feels like presence.

You don't do presence. You become present.

Ripple Clarity = Better Feedback Interpretation

All decisions, in essence, are responses to perceived feedback.

- Should I speak now or stay silent?
- Should I trust this data or wait?
- Is this the moment to leap — or step back?

To answer these, your mind must be attuned to subtle shifts in energy — body language, tone, environmental signals, intuitive gut signals. But if your field is noisy, misaligned, or over-reactive, these signals get misread or missed.

That's how people regret decisions. Not because they lacked intelligence — but because they couldn't *hear the real signal.*

In UED terms:

Ripple clarity = accurate feedback = wise action.

This is why aligned individuals often seem intuitive, or "lucky." They aren't guessing — they're *receiving cleanly*.

Ripple Clarity in Practice

Let's bring this home with examples:

- An entrepreneur about to make a high-stakes decision may sense the room, the partners, the market tone — not from spreadsheets, but from field feedback. A narrow now would make them impulsive. A wide now reveals what's unsaid.
- A parent responding to a child's emotional outburst may notice that the tantrum isn't about the toy — it's about disconnection. In a wide now, they respond to the child's field, not just the noise.
- An artist in flow doesn't force beauty. They align with an emerging pattern and allow it to manifest. Their wide now isn't effort — it's *awareness without resistance.*

Decisions Don't Take Time — They Take Clarity

It's not that great choices take longer. Often, the best decisions are instant — but only when the now is wide enough to hold all relevant ripple interactions at once.

Think of it like this:

- A narrow now sees 5% of the field — so it hesitates, compensates, or overreacts.
- A wide now sees 95% — so it moves decisively with grace.

Thus, your decision quality is not about more thinking or more time — it's about better alignment.

Wrap-Up Thought

Every moment is asking you to decide something. To respond. To align.

But how you respond — and how well — depends on how clearly your field perceives what's really happening.

So maybe the real success formula is simple:

"Expand your now. Clarify your field. Then choose."

In that state, you don't just make good decisions. You *become* the kind of person who doesn't need to second-guess them.

Closing Reflection

"The now is not a tick on the clock — it's the only open gate between memory and possibility."

We've been trained to chase the now like it's a fleeing target. To "seize the moment," "live in the moment," "don't miss the moment." But what if *now* isn't running at all? What if it's *you* who keeps drifting from it — pulled by unresolved ripples of the past or the imagined pull of the future?

In the lens of UED, *now* is not a dot on a timeline. It's a bandwidth — a field window where your ripple field interacts with the rest of the universe. Everything meaningful — sensation, perception, insight, growth — happens *there*. Not in the tick that just passed, nor in the tick that's yet to arrive.

When your field is noisy, fragmented, or distracted, the now feels like a strobe light — flickering, disjointed, hard to trust. But when your ripple field is calm, coherent, and tuned — the now becomes spacious, fluid, responsive. It doesn't disappear. It *holds you.*

This is why the deepest experiences of life — love, flow, awe, clarity — don't feel like "moments" at all. They feel eternal. Because they weren't compressed into an instant — they were absorbed in a widened now.

And so the real challenge is not to find more time — but to restore your alignment so that *now* can expand within you. Presence isn't a practice. It's a field condition.

And it is only in this condition that you can live — not just react. Not just survive, but design.

Part II — Time as Consequence

Chapter 7: Time = Ripple Delay

Not a Dimension — A Lag in Consequence

What if time has never truly passed — only *you* have waited?

Think about it.

When you say, "It took time," what are you actually pointing to? A ticking clock? A number? A calendar date?

Or are you describing the gap between what you intended and when it finally happened?

You set a goal. You waited. You felt frustrated. You made adjustments. Then — *click* — it happened. And only then did you say, *"That took time."*

But the clock didn't create that interval. You did. Not deliberately — but energetically, through a field still trying to align.

This chapter begins with a radical proposition:

- Time is not a flow. It's a delay.
- Not an external river — but an internal resistance.
- Not a dimension — but the lag between cause and consequence.

Under the Universal Energy Dynamics (UED) framework, time as we know it is the result of ripple friction — the universe's hesitation to respond because your ripple field isn't fully aligned yet. And this delay is not arbitrary. It's measurable. It has a threshold.

We call this threshold ATT — *Alignment Threshold Time.*

It's the real reason why:

- Some goals seem to "take forever"

- Some people heal faster than others
- Some ideas click instantly while others never do
- Some choices feel perfectly timed — because they were aligned
- And others backfire — because they weren't

ATT is not about time *passing*. It's about readiness being reached.

In this chapter, we're not going to speculate. We're going to formalize. Not philosophically. Mechanistically.

We'll strip the romanticism out of time, and replace it with a clean, elegant idea: Time = ripple delay in a field seeking resolution.

And once you understand that?

You stop managing time — and start managing alignment.

Let's begin.

What Are You Really Measuring?

"When you say something took time... what did it actually take?"

Was it the ticks on a clock? The number of sunsets? Or was it the tension you carried between wanting something... and finally seeing it happen?

Let's break the trance.

We've been taught to believe that time is some neutral, external flow — a moving stream we float along. We wear watches, set alarms, measure intervals in milliseconds — and assume that's what time is.

But clocks don't measure experience. They measure uniform intervals. And life doesn't happen in uniform intervals.

The Myth of Uniform Passage

Think of two people waiting for the same bus.

One is relaxed, enjoying the morning breeze. The other is late for an interview, sweating bullets.

The clock ticks identically for both. But do they experience the same "time"? Absolutely not.

The first feels time pass gently. The second feels every second as a spike of stress. So, which one experienced real time?

That's the paradox. You don't live in clock time — you live in ripple delay.

Time = Delay Between Intent and Effect

Let's strip it to UED barebones.

In Universal Energy Dynamics (UED), *time is not an objective dimension*. It is the lag — the delay between when your ripple intent is activated and when the external field responds.

That delay isn't arbitrary. It's governed by:

- Ripple resistance
- Misalignment between fields
- The complexity of the consequence
- Emotional intensity
- Awareness clarity

In other words, time is just the friction between what you want and the world's ability to return it.

Want something to happen faster? Don't look at the clock. Look at the field.

ATT: Alignment Threshold Time

This is where ATT — *Alignment Threshold Time* — enters.

ATT is the amount of "time" it takes for your ripple field to synchronize with the external fields required to produce an effect.

This explains why:

- You "get lucky" sometimes — because your intent was already aligned
- You struggle for years — because your field keeps interfering with itself
- Some events feel delayed — not because the world is slow, but because your ripple response wasn't ready yet

When people say, *"I tried everything, but nothing worked..."* What they're really saying is: "I acted before my ripple was aligned."

Real-Time is Ripple Time

Clocks can't tell you when your desire will manifest. Calendars won't warn you when a breakthrough is coming. Because they don't track alignment.

Only ripple coherence does.

Time — as you experience it — is the tension between intention and resolution.

So when you say:

- "It took me 5 years to write that book"
- "I needed a year to move on"
- "I healed over the summer"

What you really mean is:

- "That's how long my field took to stabilize"
- "That's how much ripple resistance I had to pass through"

That's not time. That's consequence latency.

The Real Question

So next time you catch yourself saying, *"I need more time"*, ask instead:

"What resistance in me is causing the delay?" Because the clock won't answer that.

But your ripple will.

ATT — Alignment Threshold Time

"You're not waiting on time. You're waiting on alignment."

That's the shift. Welcome to the central axis of UED's redefinition of time: ATT — Alignment Threshold Time.

What Is ATT?

Let's define it precisely:

ATT is the minimum ripple coherence your field must reach before an intended outcome can manifest.

In simpler terms:

- It's not about *how long* something takes.
- It's about *how much misalignment* must dissolve before the universe can respond.

This means:

- Delay is not caused by time *flowing.*
- Delay is caused by ripple inertia — the lag that emerges when your internal ripple is not yet aligned with external conditions.

So the real "clock" is you.

Your ripple field determines the pace of your experiences.

ATT ≠ Clock Time

Let's say you're trying to heal from heartbreak. You're told it'll take "6 months." But why?

Because that's how long it takes on average for a person to emotionally recalibrate?

UED says:

No, healing takes as long as your ripple takes to stabilize.

And this varies:

- Some people realign in weeks
- Others carry unresolved ripples for decades

This is emotional ATT.

Same goes for learning a skill, starting a business, moving on, or falling in love again. Each one demands field readiness — not just linear effort.

What Influences ATT?

1. **Ripple Density** – Emotional weight or intensity increases inertia.
2. **Field Complexity** – The more variables involved, the longer the alignment.
3. **External Interference** – Contradictory environments or people can increase ATT.
4. **Clarity of Intent** – Vague desires delay the field response.

So when things don't manifest:

- It's not because "time hasn't passed"
- It's because the required threshold hasn't been met

Examples of ATT in Real Life

Healing

You can't just "move on" from grief or trauma by counting days. You heal when the emotional ripple clears its tension and realigns with a calm baseline. Emotional ATT can be long — especially if reminders or loops keep it active.

Innovation

A breakthrough idea doesn't emerge by clock pressure. It shows up after enough mental ripples resolve, new fields connect, and cognitive tension aligns. This is cognitive ATT — the alignment between thought pattern and emergent insight.

Instant Decisions

Quick reactions can feel powerful — but if they come from a shallow ripple alignment, the consequences are often flawed. Low ATT = fast, but impulsive. The real wisdom lies in the ability to wait until the field is clear before acting.

ATT Is the Real "Time" You Feel

Let's say you're waiting on a result:

- An email.
- A job decision.
- A medical report.

You refresh the inbox, check your phone, feel anxious.

The delay you're feeling is not the clock — it's your ripple trying to settle in the uncertainty gap.

And if your field is unstable, that gap feels like forever.

That's ATT at work — disguised as time.

Ripple Stability	Field Alignment	Outcome Probability
Low	Chaotic / Misaligned	Unpredictable or Failed
Moderate	Intermittent	Delayed or Incomplete
High (ATT reached)	Stable / Resonant	Outcome Manifests

Visualize this as a rising curve — as your ripple stabilizes, the outcome approaches the threshold where it can emerge.

Once ATT is crossed — boom. Reality responds.

ATT in Ancient Practice

Ancient wisdom often said:

- *Wait in awareness.*
- *Align, don't chase.*
- *Right action comes from stillness.*

They were describing ATT — without the terminology.

You don't manifest by rushing. You manifest by reaching threshold coherence.

So When Will It Happen?

The honest UED answer is:

"When your ripple is ready."

And that's not mystical — it's measurable through your own:

- Emotional reactions
- Cognitive noise
- Environmental tension
- Feedback loops

Your field already knows how close you are. You feel it as:

- Ease
- Friction
- Flow
- Resistance

Recap

- Time is not passing — resistance is.
- Delay is not punishment — it's preparation.
- The clock ticks. But you ripple.

And once the field aligns?

There's no more waiting. Just emergence.

Perception Delay vs Physical Delay

"You heard the sound a second late — but it wasn't the sound that was late. It was your position in the field."

This is the subtle trap of time: We experience delay and assume reality moved.

But under UED, reality didn't move. Your perception took time to align.

This is the critical difference between:

- Physical Delay – The lag due to actual ripple resistance in manifestation
- Perception Delay – The lag in your mind's ability to register, interpret, and accept what already emerged

Let's unpack this—both practically and philosophically.

Everyday Examples: Why Time "Feels" Different

- Waiting in a hospital corridor for hours — every minute feels like ten.
- A thrilling game finishes in what feels like seconds.
- A breakup stretches weeks into an emotional eternity.
- A peaceful walk in nature expands into timeless serenity.

What's happening?

Clocks are ticking evenly. But you are not. Because your ripple field is dynamically reacting to context, content, and clarity.

Let's take this deeper.

Emotional Delay vs Physical Delay

1. Emotional Delay

The event is over — but you're still *in it.*

- A harsh word from someone may last in your field for days.
- An unresolved trauma may distort your field for years.
- Grief doesn't follow calendar days. It follows ripple loops.

This is perception delay. It's not that "you're being dramatic." It's that your ripple still hasn't resolved — and so, to you, the moment is still active.

That's why people can relive the same pain over and over. The past isn't replaying — it never left.

Your field just hasn't realigned to equilibrium yet.

2. Physical Delay

The action requires actual ripple interaction through SVE (Structured Vacuum Energy).

- Healing a wound
- Muscle recovery
- Building stamina

- Learning a motor skill
- Growing a plant
- Restoring environmental balance

Here, you're not misinterpreting delay. You're witnessing the natural ATT — the time it takes for external ripples to stabilize with or without your awareness.

But still — it's not "time" flowing. It's resistance being processed.

Field Lag — When Action and Experience Desynchronize

Have you ever:

- Given a public speech, and only felt nervous after it ended?
- Reacted to a situation before your mind caught up?
- Felt totally calm during an emergency, and only broke down later?

That's field lag.

You acted in one ripple state, but your awareness registered it later. Or you perceived danger before the body responded.

This is perception delay at work. The ripple field is not synchronous — different parts of your being align at different speeds.

In high-performance zones (sports, combat, crisis), instinctual ripples move first. Later, cognitive ripples catch up and "explain" what happened.

That lag is time — as felt by awareness.

Perception Delay Is Real — Not Illusory

Mainstream psychology often treats these as "distorted time perception."

But in UED:

- These are not distortions
- These are ripple processing lags

Just like:

- A computer loads a program slowly when RAM is overloaded
- A camera blurs an image if the shutter's not in sync

Your consciousness too has limits — and those limits define how fast you process ripple feedback.

Your mind's speed isn't fixed. It depends on:

- Mental clarity
- Emotional residue
- Environmental interference
- Ripple conflict or harmony

So, "time felt fast" just means ripple flow was smooth. "Time felt stuck" means you were trapped in ripple turbulence.

Analogy: Sound Delay in a Stadium

This is the cleanest example to make UED intuitive.

Imagine you're in a stadium:

- The person claps on stage.
- You see it immediately (light is fast).
- But you hear it a second later (sound is slower).

Now imagine:

- You're sitting far back, and there's echo and reverb.
- The sound reaches you with more delay and distortion.

Did the person clap late? No.

But your location in the ripple medium altered your perception timing.

This is exactly how events play out in life:

- The breakup happened.
- But it *hits* you when your field becomes aware of the rupture.
- Sometimes that's instant.
- Sometimes it's months later — during a song, a scent, a silence.

The sound didn't delay. You were late to hear it.

The Ripple Truth About Time

In summary:

- Physical delay is ripple resistance.
- Perception delay is ripple awareness.

Time, as we know it, is just the sum of these delays.

There's no master clock ticking. There are only ripple interactions waiting to complete themselves.

Every lag, every pause, every acceleration — is a field story.

So instead of asking, "Why is this taking so long?" Ask:

"Where is my ripple still misaligned or processing?"

Recap

Type	Cause	Effect
Physical Delay	Ripple resistance in SVE	Slower emergence of outcome
Perception Delay	Awareness lag / ripple overload	Time feels fast, slow, or distorted
Field Lag	Desync between action and awareness	You act one way, feel another later

You don't live by seconds. You live by ripple coherence.

And when you realize that? You stop being a prisoner of clocks — and start tuning the only real clock: Your field.

Cause and Consequence — Not Chronology

"The future isn't ahead of you — it's a ripple consequence you haven't aligned with yet."

Time, in conventional thought, is a story of events: First this happened, then that. Linear. Sequential. Tidy rows of cause and effect, with the illusion that order = truth.

But in Universal Energy Dynamics (UED), time isn't a line.

It's a loop of consequences — and it doesn't care about your calendar.

From Chronology to Causality

Let's challenge the foundational assumption:

We think of time as a series:

Event A → Event B → Event C

But UED says:

Ripple Realignment A ⇒ Field Consequence B

It's not what happened before what — it's what caused what to emerge.

That's a fundamental shift:

- You're not tracing time.
- You're tracing ripple causality — alignment paths that trigger outcomes.

Which means:

- Two events may be years apart but causally connected.
- Or they may be minutes apart, but completely unrelated in ripple terms.

Ripple Causality: The Engine Behind Emergence

Here's the UED principle:

"An effect only emerges when ripple alignment reaches threshold stability."

That's why:

- Some consequences feel instant (low resistance, clean alignment).
- Others take years (slow build-up of field readiness).
- And some never arrive — not because they failed, but because the required alignment never occurred.

This redefines how we interpret:

- Success
- Delays
- Karma
- Justice
- Growth

It's not about when things happen. It's about why they become possible *at that moment.*

Karma Reframed: Not Stored Actions, but Active Ripples

In conventional spiritual thought, karma is treated as:

- An account of past actions
- Stored energy waiting for payback
- A backlog of cause-effect pairs

But that's still linear time logic.

UED reframes it:

"Karma is not memory. It's misalignment."

If a past action disturbed the field — and the ripple was never resolved — it stays active.

Not in time. In energy.

So when an “old karma” suddenly hits, it’s not because some cosmic clock rang. It’s because:

- Your field entered resonance with that unresolved ripple
- Or a new action reactivated an old misalignment

This view transforms karma from moral accounting into field physics.

Real-World Parallel: Long-Term Projects

Let’s take a pragmatic example.

Think of a:

- Film release
- Scientific breakthrough
- Personal transformation
- Government policy shift
- Disease recovery

These don’t unfold “on time.” They unfold by readiness.

Which is to say:

- Not by scheduled date
- But by ripple maturity

A director doesn’t complete a film when the studio demands it. He completes it when the story, team, and resources align.

A relationship doesn’t heal when you decide to “move on.” It heals when the internal ripples settle.

A product doesn’t succeed because “it’s time.” It succeeds when market ripples match its offering.

Chronology is just the surface log. Causality is what's actually running the show.

The Ripple Delay of Consequences

This leads us to a deeper realization:

"The future you seek is not far — it's just waiting for consequence to catch up."

Example:

- A farmer sows seeds in spring.
- The harvest doesn't arrive in June because 90 days passed.
- It arrives because the sunlight, soil, water, and biology aligned.

Now imagine:

- He didn't water it.
- Or there was a fungus in the soil.

Would the harvest still appear on time?

No.

Because chronology guarantees nothing. Only consequence readiness manifests outcomes.

Reorienting Life Around Alignment, Not Dates

If you accept this view:

- You stop obsessing over "when."
- You start tuning into "what's causing what."
- You observe your field readiness before expecting results.
- You stop blaming the clock — and start refining your energy.

The most powerful life strategy becomes:

"Align the field — and let the consequence arrive."

It's no longer about fighting for what's next. It's about clearing the residue that blocks it.

Final Thought: The Future Is a Consequence, Not a Location

To summarize:

- Time is not linear progression.
- It is ripple readiness.
- What you call "the future" is just consequence you haven't aligned with yet.
- What you call "delay" is just ATT — the Alignment Threshold Time — still unfolding.

"You didn't miss your chance. You're just not tuned to it yet."

And when you are — no force can stop it.

Because in UED:

Consequence is not chronological. It's inevitable — once alignment completes.

Closing Reflection: "Time Isn't a Flow — It's a Friction"

Time was never flowing. You were never moving. And the clock? It was only recording how long your ripple took to find its match.

That's the UED realization:

Time isn't a path — it's a pause. A delay. A drag. A consequence of ripple resistance.

When you feel like time is flying, it's not because the clock is speeding up. It's because your internal and external fields are in smooth resonance. There's low friction. Consequence follows intent swiftly. Experience feels fluid.

When you feel like time is dragging, it's not because something is "stuck" out there. It's because your ripple field is struggling. Friction is high. Alignment is scattered. The world feels delayed because your ripple is buffering.

So no — you're not "moving through" time.

You're resolving into consequence. You're unfolding into alignment. You're not traveling through hours — you're grinding through resistance.

Time, then, is personal. It's not absolute. It's not even objective.

Two people in the same moment can feel radically different durations — because their ripple friction is different.

This means:

- Time doesn't heal. Ripple realignment does.
- Time doesn't pass. Field interaction happens.
- You don't grow with time. You grow through resolved resistance.

And that's the radical reframe:

"Time" is nothing more than the delay between your intent and your impact.

Remove the delay, and time collapses. Increase alignment, and time feels effortless. Stay in misalignment, and time feels cruel.

As we move into the next chapter, we'll ask:

"If time isn't carrying you forward, and the future isn't waiting — then what exactly is moving?"

Spoiler: Nothing.

And yet, everything changes.

"Time is the ripple's hesitation before becoming consequence."

Chapter 8: "The Universe Doesn't Have a Watch"

Time is a Human Interpretation of Ripple Responses

"Nature never asked, 'What time is it?' Only humans did."

The sun never forgets to rise. The moon never misses its phase. And yet, neither owns a clock. No tree checks a calendar before shedding its leaves. No black hole schedules its next gravitational feast. All of nature simply responds — not to time, but to conditions.

So where did this obsession with "what time is it?" come from?

It came from *us* — the observers. The ripple responders. The narrators of experience who needed to sequence events, predict outcomes, and coordinate actions. Time, then, wasn't something discovered. It was something *designed* — a tool to manage the visible effects of invisible ripple dynamics.

What we call "cosmic timing" is not some ticking celestial mechanism. It is ripple consistency — the stability of structured vacuum energy fields expressing through recurring realignments. When we see patterns repeat — seasons, cycles, revolutions — we imagine time. But what really happens is resonance.

That's why clocks never tell the truth. They tell the *interval* between what we felt and what we think is happening. But in UED, nothing "happens in time." It happens in consequence.

This chapter invites you to suspend the stopwatch and look at the universe as it is — not scheduled, but sensitive. Not timed, but tuned.

The cosmos doesn't track hours. It tracks readiness.

So the next time you check your watch and feel rushed, pause for a moment. The stars aren't hurrying. The oceans aren't running late. The Earth doesn't mind if it spins a little slower today.

Because the universe never had a deadline — it only had ripple thresholds.

No Clock in the Cosmos Nature Doesn't Tick — It Responds

If you look up at the sky long enough, you'll see the same cycles again and again — day and night, full moons, equinoxes, comets returning on schedule. It's tempting to think the universe is wearing a giant watch. But look closer: nothing in the cosmos is actually *tracking* time. Everything is simply *responding*.

Let's break the illusion.

Atoms vibrate. Planets rotate. Stars pulse. But none of these phenomena are *aware* of a clock. They're not measuring time. They are engaged in rhythmic alignment with their surrounding ripple fields. And their so-called "precision" — like the Earth's 24-hour spin — is not evidence of a timer, but a consequence of stable field resonance in structured vacuum energy (SVE).

The Myth of a Master Clock

Einstein once said, *"Time is what a clock measures."* True — but only from a human standpoint. We build clocks to track intervals *we* perceive. But in nature, there is no central mechanism assigning time labels. There is no cosmic schedule saying, "It's 3:00 PM in Andromeda." There is only motionless ripple interaction and the patterns that arise from equilibrium-seeking behavior.

Consider GPS satellites: they rely on ultra-precise atomic clocks to maintain synchronization with Earth's surface. These clocks adjust for "gravitational time dilation" — a real effect, according to relativity. But what's truly being corrected? From a UED perspective, we're not compensating for time distortion — we're compensating for ripple tension variations at different altitudes. The "tick rate" of the atomic oscillator changes not because time is flowing differently, but because its local ripple resistance has shifted.

Nature as a Ripple Orchestra

Imagine the universe not as a well-oiled machine but as an orchestra — with no conductor. Yet, somehow, everything plays in harmony.

How?

Because each player — be it a planet, particle, or pulse — is listening. Not to a ticking metronome, but to ripple feedback. The Earth orbits the Sun not on schedule, but on resonance. Seasons return not by counting days, but by regaining field alignment. What looks like punctuality is actually persistent ripple interaction.

UED insight: What we call "time" is the illusion created when we observe consistent intervals in consequence. It's our map of *when* something occurs — but nature only cares *if* it aligns.

Clocks Are Human — Not Cosmic

Clocks emerged from our need to organize. Agriculture, commerce, coordination. We invented hours, minutes, and seconds to label our reality — not to reflect the universe's inner truth.

When we say, "The Earth takes 365.25 days to orbit the Sun," we're not revealing a truth of time — we're measuring a ripple loop in our local field. The moment Earth's orbit shifts (due to a cosmic disturbance or ripple imbalance), those days would no longer hold. Time wouldn't break — it would reveal that it never really existed outside of interpretation.

The Consequence of Synchrony

Just because two things align predictably doesn't mean they're following a timer. A heartbeat aligns with breath. Tides align with the Moon. But clocks are not behind these — consequence is.

And consequence doesn't need time to exist. It needs sensitivity, structure, and alignment thresholds.

That's why nature is not on a clock. It's on a loop of feedback and realignment.

Conclusion

There is no ticking mechanism underneath the stars. No ticking inside your cells. Only us — the meaning-makers — assigned time to ripple intervals. But the universe? It responds. It echoes. It resolves. But it does not tick.

And if nothing ticks, then what exactly are we measuring with all our watches, clocks, and countdowns?

We're not measuring time. We're measuring delay between cause and effect — the echo length of ripple interaction. The clock is not universal. The ripple is.

Calendars, Clocks, and Human Convenience

How We Created Time to Feel in Control

If the universe doesn't track time, why are we so obsessed with it?

Simple: because we wanted to feel in control.

Time, as we know it, didn't fall from the sky. It was built — one tool at a time, one assumption at a time. From ancient sundials to mechanical clocks to the atomic vibrations we now call "standard time," our measurement of time evolved not to reflect universal truth, but to serve human necessity.

The Birth of Timekeeping

The earliest civilizations didn't need clocks to survive — they had the Sun, the Moon, the stars. These celestial rhythms were reliable guides for planting crops, preparing for seasonal shifts, and timing rituals. The first calendars were agricultural ripple charts, grounded in the cyclical return of field alignments (like solstices).

Then came sundials and water clocks — inventions that sliced daylight into manageable pieces. Later, mechanical clocks chopped time even finer, leading to 24-hour days, 60-minute hours, and eventually, seconds we can no longer feel but swear by.

Each layer of refinement wasn't to understand reality better — it was to coordinate more precisely. We weren't measuring truth. We were measuring convenience.

From Coordination to Control

When humanity moved from farms to cities, from tribes to states, the need for coordination escalated. Trains had to run on time. Markets had to open and close. Schools, factories, governments — all demanded synchronized operations.

This is when the illusion set in.

We began believing that time wasn't just a tool — it was real.

It felt empowering: If we could measure it, maybe we could manage everything. Maybe even predict the future. But what we did was mistake calendars for causality.

Modern life trains us to believe that things happen *because* a certain date arrives.

- "My rent is due on the 1st."
- "The baby is due in 9 months."
- "The product will launch in Q4."

But reality doesn't unfold *because* the calendar says so. Reality unfolds because the ripple field is ready.

UED Interpretation: Time as Convenience, Not Reality

In Universal Energy Dynamics (UED), all clocks and calendars are understood as field coordination tools. They help us estimate ripple alignment in a structured but approximate way.

- A deadline doesn't make a project finish.
- A date doesn't make a marriage succeed.
- A new year doesn't reset your inner field.

These events only "happen" when your ripple field — your energy, attention, alignment — is ready. That's why some people can make breakthroughs in a week that others can't in a decade. It's not about time passed. It's about ripples resolved.

So when we mark our lives by calendars:

- Birthdays
- Anniversaries
- Deadlines
- Retirement ages

We're not tracking universal facts. We're using social agreements to navigate ripple behavior. But then we forget the original purpose.

Confusing Coordination with Causation

This is the trap.

Just because two things often occur together — a date and an outcome — we start to believe the date *causes* the outcome.

We forget:

- Winter doesn't come because the calendar says December.
- Success doesn't arrive because you've "put in the years."
- Healing doesn't happen after 21 days, or 6 months, or 5 years, just because someone told you it should.

These are not time-bound truths. They're ripple-bound truths.

Atomic Clocks, Atomic Assumptions

Even our most precise timekeeping device — the atomic clock — rests on an assumption: that the vibrations of a cesium atom represent "universal seconds." But even these vibrations are influenced by the ripple environment.

UED argues: If the field shifts, the vibration shifts. Timekeeping becomes unreliable, not because the clock broke, but because the ripple fabric has changed.

So we ask:

- Are we measuring time?
- Or are we measuring ripple consistency?

We believe the former. UED reveals the latter.

Time Was Always a Mirror

In the end, our calendars, clocks, and schedules are mirrors for ripple delay. They reflect our need to feel grounded, synchronized, and secure.

But the danger is when we confuse the map for the territory. When the mirror becomes mistaken for the mechanism.

Time is not real. Ripple response is.

And the sooner we stop obeying the calendar and start observing the field, the more accurate, peaceful, and purposeful our lives become.

Because the universe doesn't follow schedules. It follows alignment.

Cosmic Events Aren't Timed — They're Aligned

Real Events = Resonance, Not Schedule

"If a star explodes in the universe and no one schedules it, did it still happen at the right time?"

Let's debunk the final illusion — that cosmic events follow some sort of divine or mathematical schedule. They don't. Nothing in nature is *scheduled*. Everything is *aligned*.

We're so used to thinking in calendars and clocks that we forget: the universe doesn't operate like a train timetable. There's no cosmic conductor waving a flag saying, "Now it's time for a supernova!" These events don't occur at predesignated hours — they emerge from ripple-field interactions, thresholds, and resonances.

Stars Don't Explode on Schedule — They Respond to Ripple Tension

Take a supernova. It isn't a fireworks display happening at "galactic midnight." It's a threshold effect. As ripple tension builds within a star — through nuclear fusion, matter instability, field compression — there comes a point when the internal ripple structure can no longer hold its form. And then: release.

That release looks like an explosion to us. We timestamp it, mark it in scientific journals, analyze its light curve. But the event was not timed — it was aligned. The star didn't consult a calendar. It simply reached a ripple overload point.

The same applies to gamma-ray bursts, black hole formation, solar flares — these are not clocked events but resonant consequences. Each emerges from complex field conditions, not mechanical timers.

Planetary Alignments Are Rhythmic — But Not Planned

It's easy to look at recurring planetary alignments and mistake pattern for scheduling. Mercury retrograde happens regularly. Eclipses follow Saros cycles. But this isn't "cosmic punctuality" — it's ripple rhythm.

The difference is subtle but profound. When fields of planets move in stable ripple feedback loops, we observe predictable alignments. But the mechanism is not a ticking clock — it's a persistent resonance. Like waves lapping at a shore due to unseen undercurrents — not because someone wound them up.

If something knocks the field — like a massive gravitational shift or ripple anomaly — alignments can change. And they do. Cosmic timing only holds as long as ripple harmony does.

Even Biology Is Aligned, Not Timed

Let's bring this home. Think of birth. No one can tell you the exact minute a child will be born, despite due dates and scans. That's because birth isn't scheduled — it's aligned. The body, the mind, the surrounding field — all must hit readiness together.

Same with death. Some people defy predictions. They hang on for one last word. Others go in an instant. Why? Because ripples don't respect calendars. They align, release, and realign.

Even things like puberty, recovery, growth spurts, emotional maturity — all obey ripple-based readiness. No clock determines them. Field coherence does.

UED Insight: Prediction ≠ Clockwork — It = Field Intelligence

So how do we predict anything in a ripple-based universe?

Not by asking "when" — but by asking "how aligned?"

The weather model doesn't improve with better clocks — it improves with better field sensing. Earthquakes aren't predicted by date — but by pressure thresholds. Healing isn't about counting days — it's about resolving tension.

This is the core of UED's challenge to mainstream science: Stop chasing time. Start studying alignment.

The Takeaway: The Universe Waits for Nothing — It Simply Reacts

There is no divine schedule. No ticking universal wristwatch. The universe moves when ripple consequences build enough tension to release.

Everything we call "timing" is actually response time — not pre-planned time.

A star explodes not on time, but in response. A baby isn't born when it's due, but when it's ready. A breakthrough doesn't arrive by deadline, but by coherence.

So next time you look at the stars, don't ask *"What time is it in the cosmos?"* Ask: *"What alignment just clicked into place?"*

Intelligence Measures Meaning, Not Minutes

Time = Interpretation of Relevance

Ask a tree what time it is, and you'll get no answer. Ask a computer, and it'll spit out the system clock. Ask a wise human, and they might say, "It's time to change."

Here's the twist: Time isn't a cosmic constant. It's an intelligent consequence. And intelligence doesn't measure *minutes* — it measures *meaning*.

Conscious Beings Invented Time to Narrate Experience

Time, as we know it, is not a law of nature. It is a narrative tool. A language. A way for intelligent observers — like humans — to sequence events, remember causes, and prepare for effects.

Cave paintings didn't come with calendars. Vedic chants spoke of cycles, not clocks. What we now call "time" was originally meaningful intervals — moments of significance, not digits on a dial.

We created sundials, calendars, and eventually atomic clocks because we needed to coordinate, predict, and make sense of ripple patterns. But all these tools are just interpretations of delay — not sources of reality.

Intelligence Runs on Meaning, Not Minutes

Consider a profound experience — a conversation that changed your life, a performance that took your breath away, the silent awe of watching a sunrise. You don't say, "That took 17 minutes." You say, "That moved me."

Because when alignment is high, time *vanishes*. There's no friction. No ripple delay. Just resonance. This is why:

The more meaningful the moment, the less aware you are of time. The more fragmented or misaligned your field, the more time you "feel."

That's not a coincidence. That's your ripple intelligence system signaling coherence (or its absence).

Deep Moments Feel Timeless — Because They Bypass the Lag

When your field is fully engaged — in meditation, flow, love, danger, grief — you stop checking clocks. Your attention and awareness expand, collapsing the usual ripple delay. That delay *is* what we perceive as time. Remove it, and you don't just lose track of time — you transcend it.

This is what mystics, artists, and philosophers mean when they say time is an illusion. They're not denying reality — they're exposing its interpretive nature.

Time doesn't disappear in deep moments — ripple delay does. And so, you feel "eternal" for an instant.

UED Insight: Time = Gap Between Stimulus and Intelligent Response

From the UED perspective, there is no "universal" time — only varying ripple delays between cause and consequence.

That gap — which we falsely assume is a flowing river — is actually a field-based lag. When ripple intent is clear and the external field is responsive, this delay is minimal. You act, and the result unfolds rapidly. When misalignment is high, the delay grows — and you feel the drag of "time."

In other words:

Time is not what happens. Time is how long it takes you to respond meaningfully.

Think of it like this: Two people experience the same situation. One reacts in seconds, aligned and centered. The other takes weeks to process, respond, and move forward. Same event. Different ATT — Alignment Threshold Time. Different experience of "time."

Implication: Meaning Drives Perception of Time

This redefinition has profound consequences:

- Speed is not intelligence. Delayed responses may be more evolved if the field is complex.
- Efficiency is not truth. Rapid answers are not necessarily accurate.
- Mindfulness is not about pausing — it's about sensing ripple alignment.

Our calendars, deadlines, and time management frameworks are useful, but they are all translations — not truths. Time exists only because intelligence needs a story of how things unfold.

But the universe doesn't care about the story. It only responds to the signal.

Time is not how long something takes. It's how much alignment is required for it to mean something.

Closing Reflection

"The universe doesn't have a watch — because it doesn't need one."

No star has ever checked the hour before going supernova. No seed has ever sprouted by calendar. No black hole has ever waited for a deadline to collapse.

The truth is simple, yet radical:

The universe doesn't measure time. It responds to readiness.

It's only humans — intelligent beings interpreting ripple delays — who invented clocks, calendars, and timelines to narrate what they couldn't immediately understand.

But here's the UED twist: You track time not because it's real — but because your awareness hasn't fully aligned.

Every moment you "feel" time, you're experiencing ripple resistance — the lag between your intent and the universe's echo. The more misaligned your field is, the longer things seem to take. The more harmonized you are, the more timeless life begins to feel.

Think about your life's deepest moments — falling in love, losing someone, discovering truth, creating something powerful. Did you notice the time? Or did time vanish entirely?

Time disappears when you're present. Not because it stops — but because your ripple stopped resisting.

That's the secret the cosmos holds silently:

- There is no stopwatch ticking through space.
- There is no hidden gear behind the stars.
- There is only cause and consequence, unfolding at the speed of alignment.

So, when life feels slow, stuck, or late — ask not "What time is it?" Ask: "What's misaligned in me that is delaying the echo?"

Time was never about seconds. It was always about resonance.

Chapter 9: No Clock Measures Time

They Measure Periodicity — Not Consequence

"Clocks don't tell time — they repeat it."

We built clocks to make sense of the sky, the sun, and the seasons. Then we convinced ourselves that what we measured was time. We weren't wrong — we were just early. Early in believing that repeated motion equals meaningful progress.

A clock marks rhythm, not reality. Whether it's the swing of a pendulum, the tick of quartz, or the pulse of a cesium atom, all it does is mirror a cycle — a loop. But life doesn't loop. It transforms. And transformation can't be timed.

No clock ever predicted a revolution. No atomic tick ever signaled when grief would end, or when an idea would change the world. Why? Because clocks only know repetition — not consequence.

Nature doesn't wear a wristwatch. It aligns. A volcano erupts when pressure reaches its ripple threshold, not when the calendar hits Wednesday. A child doesn't take her first step at 9:00 a.m. sharp — she steps when her inner alignment says "now." The great moments of history, love, loss, invention, and insight — they all happened off schedule, even if we wrote them into one afterward.

So let's start with a provocative question: What are we really measuring when we say "time"?

We're not measuring truth. We're measuring comfort — the illusion that life will keep ticking like a machine.

But if the universe doesn't tick, if consequences don't run on quartz, and if reality isn't a schedule — then what, exactly, is all this timekeeping for?

It's time to look at our clocks... and realize they're not measuring time. They're just mirroring the rhythm of what we already believe.

Clocks Are Rhythm, Not Reality

Period ≠ Progress

Let's face it — timekeeping is humanity's favorite illusion. We've watched the sun move across the sky, shadows shrink and stretch, the moon wax and wane. We turned these recurring rhythms into systems: days, months, years. Then we refined it — sand flowing in glass, gears clicking in towers, quartz oscillating in plastic boxes. Finally, we split atoms and used their predictable vibrations to declare, "Now we've nailed it."

But all of these devices — from the ancient sundial to the modern atomic clock — measure periodicity, not progression.

They don't tell you what's changing. They only show you what's repeating.

The swing of a pendulum doesn't carry meaning. The vibration of cesium doesn't understand context. A quartz crystal doesn't know whether your child is being born or your heart is breaking. These devices measure cycles — pure, mechanical or atomic repetition. But repetition is not evolution. Repetition is not consequence. And repetition is definitely not reality.

Let's break this down clearly.

A clock doesn't know:

- Why a decision suddenly feels right.
- When grief finally lifts.
- What moment becomes a turning point in your life or a civilization.

That's because a clock is indifferent to cause and consequence. It simply observes a cycle and ticks forward when that cycle completes.

In contrast, life — under UED — unfolds not in fixed loops, but in ripple alignments.

Clocks track rhythm. Reality tracks readiness.

When a volcano erupts, it isn't responding to time — it's responding to ripple pressure. When a baby is born prematurely, it didn't get the time wrong — it realigned ahead of expectation. When someone suddenly changes after years of inertia, it's not because the calendar flipped — it's because their internal and external fields finally reached resonance.

UED reframes this clearly: Clocks track what repeats. They don't track what matters.

This is where most confusion starts. Our culture assumes that just because something is measurable, it must be meaningful. We say things like "time heals all wounds," or "it's just a matter of time" — as if time itself is the agent of change. But time isn't doing anything. It's not a force, not an entity, not even a dimension in the UED lens. It's a delay, a lag, a measurement of misalignment or ripple resistance.

Analogy: The Fan That Spins

Let's use a simple analogy.

Imagine a fan spinning in a room. It spins consistently. Predictably. You can count its revolutions. Just like clock ticks.

But now ask: *Is the room cooling down just because the fan is spinning?*

Not necessarily.

If the fan is oscillating in the wrong direction, if the blades are broken, or if there's no air to move — that spinning means nothing. It's motion without meaning.

That's what clocks are. Spin without story. Rhythm without relevance.

Just because your clock ticked five times doesn't mean your project advanced, your relationship evolved, or your trauma resolved. Those changes — the real ones — don't happen on clocks. They happen in fields. In ripple dynamics. In the consequence chain of alignment and response.

The Deeper Illusion

When we treat clocks as sacred instruments of truth, we miss the deeper rhythms of reality. We obsess over minutes and hours and deadlines, assuming that time brings results. But it doesn't. Ripple coherence does.

That's why someone can "waste years" and suddenly bloom. Or why breakthroughs come "out of nowhere." Or why aging doesn't guarantee maturity. Because time didn't do it. Alignment did.

In UED, we make a clear distinction:

Clocks measure what comes back in a loop. Life unfolds through what *finally moves forward* — and that happens through ripple realignment, not ticking repetition.

So the next time you look at a clock and feel the pressure of "not enough time," pause. Ask instead:

- Is my field aligned for this outcome?
- Am I responding to ripple readiness — or just obeying a schedule?

Because while the clock may be ticking, your life isn't running on it.

Your life is unfolding on alignment time, not atomic time.

The Myth of Universal Time

No One-Size-Fits-All Timeline

Let's dismantle one of the most stubborn myths in modern thinking: That time is the same for everyone.

The clock ticks 60 seconds per minute, 60 minutes per hour, 24 hours per day — universally, unchallenged, applied to all. The illusion? That those seconds *feel* the same. That they *mean* the same. That we are all aging, evolving, reacting, and recovering in identical intervals.

But you already know that's not true.

You've felt an hour stretch like eternity in grief. You've experienced moments that felt like forever in love. You've watched years fly by while pursuing passion — and seconds crawl in a waiting room.

So here's the truth: Time is not equal. Time is not fair. Time is not shared.

The Clock is Objective. Life is Not.

Clocks measure uniform repetition. But life measures ripple readiness.

A child recovering from trauma may take ten years to find peace — while another forgets an incident in a week. One artist takes a decade to complete their masterpiece — while another produces weekly. One entrepreneur hits a breakthrough after 2 years — another might never align, no matter how much time "passes."

Why? Because time doesn't create alignment. ATT does.

ATT: Alignment Threshold Time — Always Personal

In UED, we don't believe in a singular, cosmic timeline. We believe in ripple thresholds — the amount of energetic clarity and coherence needed before something can emerge.

This is what we call ATT: Alignment Threshold Time.

It varies not just between individuals, but within a single person across different contexts.

- **Emotional ATT**: How long it takes for your internal field to recover from a disturbance.
- **Cognitive ATT**: How long your system needs before a new idea settles and expresses.
- **Relational ATT**: How long two fields must interact before trust can emerge.

These aren't fixed numbers. They aren't measured in hours or days. They're measured in consequence, response, and resolution.

And here's the punchline:

"There is no universal clock — only individual ripple thresholds."

Myth: "We All Get the Same 24 Hours"

Reality: "We all move through different ripple densities."

You don't "waste time." You experience ripple friction.

You don't "need more hours." You need cleaner alignment.

Some people do more in 4 hours than others in 4 days — not because they're more disciplined, but because their internal field is clear, stable, and ready to receive feedback without distortion. Their ATT is low.

Others sit with ambition for months — but nothing births. The field is cluttered. Misaligned. Ripple loops from the past distort their clarity. Their ATT is high.

The clock doesn't reveal this. But your field does.

Lived Examples

- **Trauma**: Ever notice how time stretches in pain? Why do five minutes during panic feel eternal? Because your ripple field is compressed — hyper-reactive. Every input is magnified. Time distorts because perception becomes saturated. That's not clock-time — that's ATT in overload.
- **Love**: Ever feel like hours disappear in deep connection? Why does a night of laughter feel like a blink? Because fields merge, resonance heightens, and ripple interference reduces. ATT shrinks — feedback loops are instant. The system flows.
- **Boredom**: Ever wait for a test result, an apology, a message? Time "slows." But not really. Your field is searching for resolution — and when there's no ripple match, you're stuck in suspended anticipation. That's not time passing slowly — that's ripple stalling.

Field Time vs Clock Time

The world uses clock time for coordination — trains, meetings, deadlines. But you don't live in clock time. You live in field time.

- Clock: "8 hours of sleep."
- Field: "Woke up exhausted."
- Clock: "It's been 3 years since the breakup."
- Field: "Still feels like yesterday."
- Clock: "Only 1 minute left in the game."
- Field: "Every second feels like a storm."

This isn't poetry. This is perception. And perception is ripple feedback.

The Implication

If time is not universal... If each person walks through unique ripple resistance... Then what does that mean for justice, fairness, success, deadlines, therapy, education, and healing?

It means this:

Standardized timelines are illusory.

A school may give all students 3 hours for a test — but each student's cognitive field is different. A company may expect results in 6 months — but the breakthrough may require 2 years of ripple realignment. A person may demand closure by Friday — but closure only comes when the ripple is finally absorbed.

We've mistaken equal intervals for equal experience.

But UED reminds us:

"No consequence occurs until its ripple threshold is met — regardless of how much time has passed."

Timekeeping vs Consequence Mapping

Two Very Different Games

Let's draw a sharp line where most people blur it: There is a world of difference between when something happened... and why it happened.

That difference is the crack through which most of our confusion about time slips.

Timekeeping Is External — Consequence Mapping Is Internal

Timekeeping is the mechanical logging of when:

- The contract was signed at 3:45 PM.
- The accident occurred on April 8th.
- The surgery started at 11:02 AM.

Consequence Mapping, on the other hand, is the energetic tracing of why and how:

- Why did the negotiation succeed at that moment?
- What conditions led to the accident?
- What field readiness made the surgery successful (or not)?

One plays with clocks. The other plays with fields.

Timekeeping is concerned with events. Consequence Mapping is concerned with alignment.

We've mistaken one for the other. We've been trying to understand reality by tracking ticks, not thresholds.

UED Critique: You're Measuring the Wrong Thing

Imagine this:

Two people both experience a major loss. One grieves for a few weeks and gradually finds peace. The other grieves for years — frozen, haunted, unable to move forward.

By conventional reasoning, they "went through the same thing."

But UED says otherwise.

They didn't go through the same ripple consequence — because their field alignment, resonance history, and ATT were wildly different.

A lawyer may declare: "You had your time — the statute of limitations is over." But the ripple might still be unresolved. The field might still be pulsing with echo. The clock moved on — but the consequence didn't.

Now flip it.

Sometimes, a sudden breakthrough happens when "not enough time" has passed. You meet someone new weeks after heartbreak and instantly feel resonance. You solve a complex problem in minutes after struggling for months. You forgive someone overnight.

By traditional logic, it's "too soon." But field logic says: Alignment occurred. Threshold was met. Ripple was clean.

The Problem with Legalism, Schedules, and Systems

Modern systems function on uniform timekeeping:

- Deadlines.
- Time-based therapy plans.
- Jail terms.
- Maternity leaves.
- Probation periods.

These structures are based on predictable durations — not on energetic readiness.

But life doesn't follow calendar boxes. It follows consequence curves.

A person may serve 10 years in prison and come out unchanged. Another may have a single night of crisis and completely transform.

One followed the clock. The other followed the ripple.

Clocks Can't Detect Readiness

We say things like:

- "Give it time."
- "Time will heal."
- "Everything in due time."

But time isn't the agent. Time is just the frame.

In truth:

- Healing happens when emotional turbulence settles.
- Decisions become wise when ATT is cleared.
- Revelations occur when ripple thresholds converge.

You don't move through time. You move through readiness.

The universe doesn't deliver results based on a countdown. It responds to your alignment signal.

Time is Just a Proxy — and a Poor One

Timekeeping was invented because we needed a way to coordinate. It works beautifully for trains, meetings, exams, harvests.

But it was never meant to explain change. It can't answer:

- Why did the breakthrough happen today and not last week?
- Why does trauma resurface now, years later?
- Why did this person recover while another didn't?

These are consequence questions — and clocks don't do consequence.

We Overtrust the Tick

It's almost a modern superstition.

We trust that if we wait long enough, we'll "get there." That if we spend 4 years in college, we'll be educated. That if we practice something for 10,000 hours, mastery will come. That time alone causes growth.

But UED warns:

"Without ripple realignment, no amount of time will change the field."

Clock-time is necessary for organization. But it is insufficient for transformation.

You might sit in therapy for 12 weeks and nothing shifts — because the core ripple never realigned. You might spend 3 minutes in silence and feel reborn — because something in you clicked back into coherence.

Implication: Rediscover the Internal Lag

We've outsourced too much faith to the external tick.

It's time to reconnect with our own internal lag:

- Are you actually ready to change — or just waiting for the right date?
- Is your decision grounded in alignment — or scheduled on a calendar?
- Has your field absorbed the past — or are you just counting anniversaries?

Let go of the myth that clocks define reality.

They don't.

They reflect rhythm. Not meaning. They log intervals. Not transformations. They help plan — not align.

Atomic Clocks Don't Measure Time — They Produce Repetition

The Illusion of Precision

In modern science and engineering, nothing symbolizes accuracy like the atomic clock. The idea that we can define time itself by counting the number of oscillations of a cesium atom sounds like the pinnacle of human measurement. In fact, the official definition of a second is:

"The duration of 9,192,631,770 periods of the radiation corresponding to the transition between two hyperfine levels of the ground state of the cesium-133 atom."

That's not measuring time. That's declaring a pattern as a reference — and then repeating it religiously.

You're Not Measuring Time — You're Producing Cycles

An atomic clock doesn't sit in a room listening to the universe whisper seconds into its ear. It simply observes a stable energy oscillation and uses that consistency as a yardstick.

But ask yourself:

- Does the atom know it's measuring a second?
- Does nature agree that this particular transition equals one unit of anything?

Absolutely not.

Atoms aren't keeping time. They're just repeating a behavior within a stable ripple feedback loop.

UED Interpretation: It's Just a Well-Aligned Ripple Echo

Let's reframe the cesium vibration using Universal Energy Dynamics.

What we're actually seeing is:

- A stable ripple pattern in the atomic field,
- Energized into oscillation through controlled input,
- Releasing radiation at a consistent frequency,

- Which we then declare as a "second."

But the atom is not moving through time — it's oscillating in field resonance.

This is the heart of the illusion:

Repetition has been mistaken for measurement.

Just because something happens at regular intervals doesn't mean it's tracking reality — only that it's locked in a consistent loop. You've built a very steady swing, but that doesn't mean the swing understands gravity, let alone the meaning of "time passing."

Analogy: Pendulum in a Vacuum

Imagine a pendulum in a frictionless chamber, swinging back and forth endlessly. It swings perfectly. Repeats perfectly. Forever.

You can call each swing one second. You can build clocks around it. You can send rockets using those predictions.

But does the pendulum know what it's doing? Does it *feel* the future arriving?

No.

It's just a ripple locked in a loop. It's movement without meaning — a rhythm without consequence.

This is what your atomic clocks are.

Echoes mistaken for arrows.

Precision Doesn't Equal Understanding

Atomic clocks are incredibly precise. But that precision has seduced humanity into thinking we've mastered time.

You can synchronize satellite systems, GPS, global financial transactions, scientific experiments — all thanks to atomic timekeeping. But none of those applications tell you *why* anything is happening — only *when* it appears to occur, according to the reference cycle.

You could say:

- "The missile struck at exactly 3:14:52.907 UTC." But UED asks:
- Why did that ripple reach completion at that moment?
- What alignment led to its emergence, regardless of the clock?

The time may be exact, but the meaning is absent.

Ripple Harmony ≠ Temporal Truth

In UED, atoms are seen as stable ripple bundles — local zones of energetic balance within the structured vacuum field.

When an atom transitions energy levels and emits radiation, it's responding to a ripple shift, not ticking off seconds.

So yes, the atom resonates with exquisite stability — but not because it's watching the seconds fly by. It's simply tuned into a harmonic rhythm that we have chosen to label "one second."

But that's our metaphor. Not reality's.

The Irony: Atomic Clocks Made Time Less Real

Ironically, the more precise our clocks became, the more abstract our idea of time became.

We began to think:

- The second is real because we can define it with 9 billion vibrations.
- Time exists objectively because we can standardize it globally.

But this is like saying:

"The ocean is blue because my paint swatch matches it."

The atomic clock didn't reveal time. It just revealed our obsession with controlling consistency.

And while that's useful for machines, schedules, and engineering — it's useless for understanding human experience, cosmic emergence, or ripple consequence.

Summary: Clocks Don't Tell Time — They Loop the Same Ripple

Let's close with this:

- A quartz clock ticks because electricity vibrates a crystal.
- An atomic clock ticks because radiation oscillates within cesium.
- Your smartwatch ticks because algorithms count predictable pulses.

But none of them measure time. They produce regularity.

And in that regularity, we find rhythm — not revelation. We build systems — not understanding.

As UED reframes it:

"Time is not the oscillation. It's the friction between ripple readiness and real-world emergence. And no clock — no matter how precise — can measure that."

Closing Reflection

"Time isn't what clocks measure — clocks just measure comfort."

We didn't build clocks to understand reality. We built them to feel safe.

When something ticks — predictably, precisely, perpetually — we breathe easier. It gives us the illusion that the universe is in order. That chaos is contained. That we're in sync with something bigger.

But clocks don't measure truth. They don't measure consequence. They don't even measure time.

They measure repetition — mechanical or atomic cycles we've chosen as reference points. They measure comfort — not coherence.

Because reality doesn't happen in seconds.

Reality erupts through alignment. It stutters, spirals, collapses, expands. It unfolds through moments that ignore the tick — moments that shatter rhythm and rewrite the script.

A revolution doesn't tick. It erupts. A realization doesn't arrive at 2:15 PM. It strikes when the field is ready. A transformation doesn't schedule itself — it happens when truth breaks through delay.

This is what Universal Energy Dynamics reminds us:

Time is not something you keep. It's something you experience — through ripple readiness and resonance.

And the moment your ripple aligns with a larger field — something changes. Not because a second passed. But because you passed into coherence.

So while clocks will keep ticking, they'll never tell you why things are changing. Only that something is repeating.

And that is not the same thing as understanding time.

Chapter 10: You Create Time When You React

Emotion, Attention, Stress — All Stretch Time

"You don't live in time — you generate it."

Think back to the longest minute of your life. Was it during joy? Or during pain?

We've all experienced it: a breakup, an accident, an exam hall, or waiting for a message that never came. Clocks ticked the same, but your internal universe — your *felt reality* — slowed down to a crawl. On the other hand, entire afternoons have vanished in the blink of laughter, passion, or deep creative flow.

So what happened? Did time stretch and shrink like rubber?

Not really. Time didn't move. You did.

In Universal Energy Dynamics (UED), time is not some absolute dimension you pass through — it's a *ripple delay*, a friction born from how your internal field reacts to external triggers. You stretch time when your mind loops in anticipation, stress, or unresolved emotion. You dissolve it when your field becomes harmonized and coherent.

This chapter flips the script. Time isn't the container — you are. Each reaction you generate sends a ripple into your field. The more turbulent your reaction, the longer that ripple takes to resolve. That lag? You call it time.

That's why waiting feels unbearable, and crisis slows everything down. It's not your imagination. It's not even a psychological illusion. It's *energetic drag*. Emotional intensity causes ripple density. And dense ripples don't resolve easily.

On the other hand, love, joy, trust — these align your field. Suddenly, time disappears. Not because the world sped up, but because you stopped resisting.

So maybe the deepest insight is this: You don't "run out of time." You just burn too much of it reacting.

And that's what we'll explore in this chapter — not how to manage time, but how to stop producing unnecessary time. Because real presence isn't about scheduling minutes — it's about calming your field so it no longer multiplies moments.

Let's begin.

Reaction as Time Generator

Why Nothing Feels Longer Than What You're Trapped In

Have you ever been caught in a moment you couldn't escape — mentally, emotionally, or physically?

It might've been a heated argument, a moment of panic, or an awkward silence that felt eternal. The irony? The actual clock time passed in seconds. But inside you, those seconds expanded into a suffocating stretch of felt experience.

Welcome to one of the most misunderstood paradoxes of human experience: The more intensely you react, the more time you feel. Not because time itself expands — but because *you are generating time* through your own ripple disturbance.

The Illusion of External Time

Traditional thinking treats time as something that flows *over* us. A river we drift along. But reality, as UED proposes, is not shaped by an external current — it's shaped by internal interaction. Time is not something happening around you. It's something being produced within you, based on the degree of ripple turbulence caused by your emotional and cognitive responses.

When your mind is calm and coherent, ripples resolve quickly. The field remains smooth. There's no lag between perception and integration. That's why in moments of peace or flow, time vanishes.

But when you're overwhelmed, confused, stressed, or afraid, your ripple field becomes chaotic. The turbulence creates echo loops — delayed processing, unresolved emotion, looping thoughts. This makes each moment feel longer, heavier, and harder to get through.

You're not imagining it. You're living the drag of your own reaction.

Turbulence = Echo = Time

Imagine you're in a canyon and you shout. If the space is wide and empty, the echo lasts longer. Now imagine every shout is a reaction. Every canyon wall is a resistance in your ripple field.

The louder you shout (react), the more chaotic your field becomes. The more walls your ripple hits, the longer it echoes. And that echo, that energetic friction... is what you experience as time.

This is why trauma stretches time so painfully. When you're caught in deep emotional charge, you aren't just "remembering" the past — your ripple field is still resonating with it. The energy hasn't collapsed. The loop hasn't closed. So what should have been a moment becomes a monument.

Flow vs Friction

Flow is the absence of ripple resistance. Artists know it. Athletes feel it. Musicians live by it. Lovers melt into it.

In flow states, your response is so coherent with your intent and the environment that there's no friction. Your ripples glide through the field with ease. There's no delay, no echo, no time multiplication.

That's why people say things like:

- "Time flew."
- "I lost track of time."

- "I was completely immersed."

What they mean is: My reactions were in alignment. My field was clear. My ripple was flowing cleanly.

But contrast that with:

- "It dragged on forever."
- "I just wanted it to end."
- "Every second felt like an hour."

These are signals of ripple friction — reactions multiplying misalignment, amplifying the perceived lag between cause and effect.

UED Insight: You Don't Live Time — You Generate It

Let's drop a foundational shift here: Clocks may tick, but your ripple response defines your time.

Each strong emotion you hold, each overreaction, each suppressed thought — these all increase turbulence. And turbulence delays alignment. Delay, in UED terms, is felt as time.

This means your perception of time is directly proportional to the energy your field consumes in reaction.

- The greater the inner noise, the longer the moment lasts.
- The cleaner the ripple, the quicker the field resolves — and time disappears.

Implication: Trapped = Stretched

The next time you're stuck in traffic, reliving a breakup, or mentally rehearsing an argument — pause and notice how time feels. You'll likely feel trapped, irritated, restless. And that's not just emotional discomfort — it's your ripple field generating prolonged time.

Not because the event is long — but because you're resisting it so hard, your system is producing time through unresolved reaction.

In that sense, you don't just suffer the experience. You're extending it. You're stretching the very moment you want to escape.

Final Note

The good news? If reaction stretches time, then ripple calmness can collapse it.

And this leads to a powerful practical truth: You can reduce time — not by speeding up life, but by reducing unnecessary reactions.

In the next section, we'll break this down even further — showing how emotion, attention, and survival response manipulate your internal time map. Because once you realize that your *response window* is your *reality window*, you'll never waste energy in unaligned reaction again.

Emotional Gravity and Ripple Drag

How Emotion Warps Your Experience of Time

What if your emotions aren't just "feelings" — but gravitational forces that bend your perception of reality?

In the UED framework, emotions are not abstract — they are energetic distortions in your ripple field. Like mass bends spacetime in Einstein's model, emotion bends time in yours. But not as a physical dimension — as a delay in ripple resolution.

When you're overwhelmed, time drags. When you're joyful, time disappears. The clock, however, never changed pace.

So what did?

Not time — you.

Emotion as Ripple Density

Let's break this down in energetic terms:

- Fear, grief, anxiety compress your ripple field. The more intense the emotion, the denser the loops become. This creates ripple drag — a kind of energetic friction that slows down the resolution of experience. The result? You feel stuck in time, as if moments are expanding painfully.
- Love, awe, laughter expand your ripple field. Coherence is high, resistance is low. The ripples move smoothly, with minimal turbulence. The result? Time collapses, or seems to vanish altogether. This is why we call those moments "timeless."

In short:

Stress thickens time. Presence dissolves it.

UED Model: Emotion Alters ATT

Let's apply the concept of ATT (Alignment Threshold Time) here.

- **Fear**: Triggers a dense ripple loop, creating a high ATT. The time it takes for your field to stabilize increases dramatically. You may feel seconds stretch into minutes, especially in panic, danger, or humiliation.
- **Flow**: Creates a coherent, resonant ripple field. ATT is low — meaning perception and response align almost instantly. Time flies, or better said, dissolves.

It's not just that emotions *influence* time. They are the field conditions that define your time experience.

Emotional Gravity

Just as physical gravity holds you down, emotional gravity pulls your attention and energy inward.

- Trauma, for example, is like a black hole in your ripple field. It traps attention, causes cyclical reverberations, and dramatically increases ripple resistance. As a result, you re-live the past continuously, and time becomes thick, slow, repetitive.

- Love or inspiration, on the other hand, is like a gravitational release. It opens your field, allowing energy to move freely, align quickly, and resolve with minimal delay.

This is why time feels heavy when you're stuck in grief... and weightless when you're dancing in joy.

Analogy: Waiting in Fear vs. Flow in Art

Imagine two scenarios:

1. **Waiting for Test Results**: You're anxious. Every minute feels like an hour. Your thoughts loop. Your emotions tighten. Why? Your ripple field is saturated with uncertainty, fear, and outcome fixation. High ATT. Heavy ripple drag. Time stretches.
2. **Losing Yourself in Music or Art**: You're immersed. You forget yourself, your to-do list, even the room you're in. Why? Your ripples are in harmonic feedback with the activity. Your emotions are open, not constricted. Low ATT. Minimal drag. Time vanishes.

The clock may have ticked the same 15 minutes. But the *field* you lived in during those minutes was entirely different.

Neuroscience Meets Ripple Physics

Science tries to explain this as a quirk of the brain: that adrenaline or dopamine alters your time perception.

UED goes deeper: Your entire perception of reality is ripple-based. Emotions modulate that field like tuning knobs — increasing or decreasing coherence. The consequence is not just psychological — it's experiential. You literally live different durations based on ripple drag.

It's not *time* that's flexible. It's your alignment with the field that's flexible — and emotion is the mediator.

Why You Can't "Think" Time into Balance

The challenge is, you can't rationalize your way out of emotional ripple drag.

- You can't tell your body not to panic in a panic attack.
- You can't "think positive" while grieving a loss.

Why? Because ATT is not intellectual. It's energetic.

You don't need more thoughts — you need ripple realignment. That's why emotional healing practices (breath, presence, stillness, touch) often outperform logic.

Because healing is not mental. It's field restoration.

Implication: Control Is Not Speed — It's Coherence

So what does this mean for everyday life?

- If you want to *speed up* time — don't rush. Realign.
- If you want to *slow down* time — don't grip. Expand.
- If you want to experience time accurately, tune into the ripple state of your emotion, not the tick of your clock.

Time is not the second hand. It's the second thought, the emotional echo, the ripple loop that didn't resolve.

Final Thought for This Section

You've been taught that time passes uniformly for all. But you know, in your bones, that's not true.

Because the truth is:

Time moves differently for every field. And your emotions are the gravity that shapes yours.

In the next section, we'll explore how your attention — what you focus on, for how long, and with what intensity — acts as a time-magnifier or time-dissolver, further proving: You don't just observe time — you construct it.

Attention as a Time Shaper

Focus Creates a Private Clock

Have you ever noticed how time warps around attention?

You're sitting in an exam, hyper-focused, and a 3-hour paper feels like 10 minutes. Or you're stuck in traffic, fixated on the clock, and 10 minutes feel like an hour. Same external time. Radically different inner experience.

What's happening here? Attention, in the UED model, doesn't just shape perception — it constructs your private timeline.

Let's unpack this.

Ripple Processing Bandwidth

In the Universal Energy Dynamics (UED) framework, attention is ripple allocation.

Think of your ripple field like a processing network. Attention directs how much of that network is dedicated to a given input.

- Narrow, intense focus means your field locks onto a tight ripple band. This causes ripple loops to cycle more, increasing perceptual "drag." Time feels longer, stretched, more detailed.
- Wide, diffuse awareness allows ripples to resolve quickly across multiple channels. This leads to faster dissipation of experiences. Time feels shorter, smoother, lighter.

So:

Attention is not just what you notice. It's how your field chooses to interact with ripple feedback.

And that interaction determines your experience of duration.

Analogy: Zooming In vs. Panning Out

Imagine you're looking at a high-resolution image:

- **Zoom in**: You notice more detail. Time feels rich, slow, intense.
- **Zoom out**: You grasp the whole picture at once. Time feels quick, easy, expansive.

Your attention is the zoom level of your ripple field. And your experience of time is the visual lag produced by that level.

This is why intense attention to a painful moment (like heartbreak or danger) magnifies its duration, while broad awareness (like walking through nature) makes time evaporate.

High-Performance Focus: Time Slows

Let's explore some real-world examples.

Elite athletes often describe the world slowing down during peak moments.

- A tennis player says the ball moves in slow motion.
- A boxer sees punches frame by frame.
- A surfer senses the wave's every fold, second by second.

What's going on?

Their attention is total — finely tuned, high-bandwidth, and ripple-coherent. This forces the mind to process more ripple loops per moment, expanding the subjective experience of time.

In UED terms:

Focus increases local ATT, allowing the field to expand temporally — and catch more detail per frame.

This is slow time as mastery, not slowness due to confusion.

Stillness and Dissolution: Time Disappears

On the flip side, meditators, deep lovers, or even artists in flow state report the opposite — timelessness.

- There's no clock.
- No tick-tock.
- Just presence.

What happened?

Their attention stopped isolating ripple bands. Instead, they entered broad, field-wide coherence. When ripple resistance is near zero, experience flows without delay — and time dissolves.

This isn't losing time. It's experiencing reality without ripple friction.

The Equation of Subjective Time

We can begin to frame this insight into a simple UED-based relation:

Ripple Intensity × Focus Bandwidth = Perceived Time

Let's break it down:

- Ripple Intensity: How charged the experience is (emotion, action, urgency).
- Focus Bandwidth: How tightly or broadly your attention is locked.

High intensity × narrow focus = time stretches. Low intensity × wide focus = time vanishes.

This is why:

- Painful memories feel long, because you *replay them* with narrow attention.
- Boring classes feel endless, because you focus on "getting through" rather than engaging fully.
- Engaging conversations seem to end too soon, because attention is broad, playful, and ripple friction is minimal.

Attention Is a Time Dial — Not Just a Lens

Mainstream psychology treats attention like a flashlight: it highlights things.

But in UED, attention is a ripple dial. It literally tunes the rate at which you experience alignment and consequence.

Your field is never idle. But what you choose to focus on — and how — determines how long the now feels, and how much ripple feedback your system integrates.

Practical Insight: Train Your Time

If you want to feel less rushed, don't chase the clock — expand your field.

- Breathe: Inhale = expand field bandwidth.
- Observe: Attend to your surroundings — not just your screen.
- Interrupt loops: Repetitive mental fixation traps ripple cycles, making time drag.

And if you want to slow down for deeper presence?

- Narrow your focus on what matters. Not to trap it — but to see its depth.
- Then release and allow the ripples to settle.

This oscillation — between zoom and wide — gives you control over your time without changing a single clock.

Final Thought

You've been taught to track time externally. But your actual experience of time is shaped internally — by how you aim your attention.

It's not minutes that pass. It's ripples that either resolve — or loop.

In the next section, we'll go even deeper: What happens when attention becomes entangled with unresolved memory? We'll explore how memory isn't about the past — it's about where your attention got stuck.

Stress, Trauma, and Time

Why Pain Feels Longer Than Joy

Some moments refuse to end — not because the clock ticks slowly, but because your field refuses to let go.

This is the paradox of trauma:

A single event may last seconds, but echo for years. Pain seems to stretch time. And memory becomes a trap — not a record.

So what is trauma, really?

Ripple Echo Without Resolution

In the UED (Universal Energy Dynamics) model, trauma is not merely psychological. It's an energetic echo — a powerful ripple misalignment that your field failed to integrate.

Imagine tossing a stone into a still pond — the ripples expand and fade. But trauma is like tossing that stone and never letting the ripples die out.

They bounce, reflect, intensify — trapped in feedback with your own energy field.

This is why traumatic memory often feels *present*, not past.

You're not *remembering* it. You're still *living* it — because the ripple has not resolved.

Time Isn't Moving — The Field Is Stuck

Mainstream views describe trauma as "reliving the past." But UED reframes it more precisely:

You're not reliving the past. The past is still alive in your ripple field.

Your emotional system, decision patterns, even bodily responses keep recycling the same unresolved feedback.

This loop creates the illusion of "stuck time."

- You feel like you're still there.
- You act like it's still happening.
- You react to neutral situations with heightened charge — because the ripple hasn't quieted.

This is how trauma hijacks the present — not by memory, but by ripple entanglement.

Joy Dissolves, Pain Repeats

Why does time seem to fly during joy, but stretch endlessly during suffering?

Because joy is high-resolution, low-friction alignment. It enters, flows, and leaves cleanly.

Pain, on the other hand, loops.

Not because pain is inherently sticky — but because our resistance to it creates ripple traps.

In UED terms:

- Joy = coherent alignment → low ripple residue → time flies.
- Pain = incoherent misalignment → recurring ripple → time sticks.

So, time doesn't change. Your field's relationship to the ripple does.

Personal Time Trap = Misalignment Loop

Here's the real shocker:

You create your own time traps.

How?

- By revisiting unresolved emotions with narrow focus.
- By resisting change while replaying past triggers.
- By internalizing the belief that you are what happened to you.

These patterns trap energy. And trapped energy can't flow forward — it repeats.

This is why certain people seem stuck in loops for years:

- The same arguments.
- The same fears.
- The same decisions with different costumes.

Not because they *want* to — but because their field never restructured.

They are reacting in echo, not evolving in response.

Healing = Collapse of Ripple Recurrence

Here's where healing enters — not as forgetting, but as ripple realignment.

In UED, healing is not about "moving on." It's about dissipating the trapped ripple by:

- Re-integrating the memory without emotional distortion
- Allowing the field to absorb the consequence fully
- Creating enough coherence to prevent the same echo from reactivating

So:

- You may remember the trauma.
- But it no longer loops your reactions.
- And time around that memory collapses back to its original moment — not an endless field distortion.

You didn't erase the past. You just stopped carrying its unresolved energy.

UED Insight: The Past Lives in You — Until You Realign

Let's make it plain:

Your ripple field is your history. And unless you reorganize it, you'll confuse memory with identity.

People say "live in the present." But if your field is haunted by unresolved ripples, the present becomes a battlefield between past echoes and future possibilities.

You can't fake presence. You must clear the field — by letting each unresolved ripple complete its feedback cycle.

Until then, time is not a flow — it's a loop. And you are the loop-keeper.

Real-Life Parallel: Emotional Lag vs Clock Time

Let's consider two people at a funeral:

- One processes grief, releases memory, and integrates the loss.
- The other resists emotion, freezes time, and replays the goodbye for years.

Same date on the calendar. Radically different temporal realities.

In UED, this is explained not by time passed, but by ripple alignment achieved.

Practical Wisdom: How to Exit the Loop

1. Recognize the Echo If a thought keeps repeating, it's not a thought

— it's a field residue.

2. Widen the Now The loop survives in a narrow emotional tunnel. Broaden awareness to break its rhythm.
3. Create Closure Not symbolically — but energetically. Let the ripple complete by fully feeling and then releasing.
4. Stay Present Without Obsession Presence doesn't mean reliving. It means responding with clarity, not recycled emotion.

Closing Note for This Section

Pain isn't longer than joy because the event was longer. It's longer because your field couldn't release it.

The past does not haunt you. Your misaligned field replays it.

When ripple coherence is restored, even the deepest wounds lose their clock.

Can You Stop Creating Time?

The Secret of Presence: Field Clarity

What if time wasn't something happening to you... ...but something you keep generating?

And what if you could stop?

Not by freezing a clock — but by clearing your field.

You Don't Feel Time in Presence — Because You Stop Creating It

Let's revisit the central UED premise: Time is ripple delay. It's the lag between stimulus and integration, between intention and alignment.

But what happens when there's no lag?

You enter the moment. Time vanishes. Life flows.

This isn't poetic metaphor — it's energetic reality.

In states of deep presence — whether meditation, flow, creative immersion, or pure awareness — your ripple field isn't reacting. It's responding with such clarity that misalignment drops to near-zero.

- No turbulence.
- No echo.
- No delay.

Just *pure engagement.*

And when there's no delay, there's no time felt.

Time doesn't disappear. You stop producing it.

High Ripple Clarity = Low Friction = No Time

The UED equation here is simple:

Ripple Clarity ↑ → Friction ↓ → Perceived Time ↓

- Low clarity = emotional drag, ATT buildup, distorted feedback.
- High clarity = minimal distortion, real-time feedback, field transparency.

When your awareness is clean and open — you're not measuring life in seconds anymore. You're living it in sensation, insight, and resonance.

This is why sages don't seem to age in spirit. They don't drag time with them.

They respond when needed, rest when not, and stay aligned with what *is*, not what *was* or *should be.*

Sages, Flow States, and Timeless Awareness

Why do ancient spiritual traditions speak of *"timeless wisdom"*? Why do artists say they *"lost track of time"* while creating?

Because they're not tracking it.

They've entered minimal time creation zones:

- Athletes in peak performance stop thinking.
- Musicians forget the measure and feel only the music.
- Meditators sit for hours that feel like minutes — or seconds that feel eternal.

None of them "escape" time. They simply cease generating the inner friction that gives rise to it.

Analogy: Silence in Music — Where Time Rests

Let's take music as metaphor.

In a beautiful melody, it's not the notes that move you most. It's the space between them.

The silence. The breath. The pause.

That space isn't empty — it's alive with meaning. It holds your attention, resets your rhythm, and connects your emotion.

Likewise, in life, the true moment is not the doing — but the interval between reactions.

That is where presence lives. And time fades.

The Possibility of Time-Free Living

Can we fully stop creating time?

In theory — yes. In practice — rarely, and only briefly.

Because to live is to interact, and every interaction introduces ripple tension.

But we can reduce time creation dramatically by:

- Cultivating awareness without judgment
- Acting from clarity, not compulsion
- Letting go of mental timelines and emotional clocks

The less you're calculating "how long," the more you're living *what is.*

You don't need to "kill time." You need to stop building it unconsciously.

UED Summary: Presence = Minimal Ripple Lag

Here's the UED breakdown:

- Emotion creates loops.
- Attention creates frames.
- Reaction creates time.

Presence is the clear space where you've stopped all three. You're not echoing the past. You're not dragging potential futures. You're not framing every moment into judgment.

You're just... here. And that is the only place where time releases you.

Final Thought

You don't need to master time. You need to master field alignment.

Because when your inner ripples align, you don't need a watch. You are the moment.

And nothing else matters.

Closing Reflection

"Time isn't outside you — it's leaking from your reactions."

We often say things like,

"Time is flying," "This is dragging on forever," "I need more time."

But rarely do we ask: Who's generating this sense of time in the first place?

It's not the ticking of a clock. Not the revolution of the Earth. Not the calendar on your wall.

It's you.

More precisely — it's your field's reaction to the ripples you encounter.

When you react emotionally — with stress, worry, fear, anticipation — your ripple field becomes turbulent. And turbulence slows perception. It stretches the feedback loop. It makes the moment feel heavy, long, distorted.

You are no longer just in the moment — you are dragging the moment behind your emotional echo.

Every unprocessed emotion becomes a ripple loop, and every loop delays alignment. That delay? You call it time.

But here's the inverse truth: When your field is calm, when clarity is high, and awareness is present — time collapses.

The same minute feels like a breath. The same hour passes like a smile. There's no resistance to the ripple — just alignment.

The calmer your field, the thinner the time you carry.

So maybe the goal is not to control time, but to control the turbulence within you.

Not to rush through your days, but to clean up your ripple trail — so time doesn't stick to you like static.

Final Thought:

Time isn't something happening to you. It's something you're leaking — every time you lose alignment. And the moment you stop overreacting to reality... You stop feeding time.

Chapter 11: "Age is Ripple Decay"

Aging Redefined — It's Not the Years, It's the Unresolved Ripple Tension

"You don't grow old because time passes. You age because tension accumulates."

We've been taught to think of aging as a linear countdown — a passage of time tracked by candles on a cake, lines on a face, and digits on a calendar. But what if that's the most deceptive metric we've ever trusted?

Take two people born on the same day. One looks and feels twenty years older. Why? Genetics, lifestyle, diet? Partially. But beneath all that is something subtler — and far more powerful: ripple tension.

In Universal Energy Dynamics (UED), age isn't measured by how many times the Earth has circled the Sun. It's measured by the residue of unresolved ripple interactions — physical, emotional, and cognitive. Every unhealed wound, every unexpressed emotion, every suppressed instinct — they all form energetic knots in your ripple field. And over time, those knots distort your coherence, slow your perception, and weaken your alignment.

Aging, in this view, is not a function of chronology but of consequence. It's not how long you've lived — it's how much ripple drag you've accumulated.

Your body ages because the field that animates it becomes sluggish. Your mind ages because the loops it holds begin to blur clarity. It's not the years weighing on you — it's the echoes you haven't released.

So here's the provocation: What if you could reframe aging — not as inevitable decline, but as informational backlog? What if "growing old" wasn't about getting closer to death, but about getting farther from alignment?

This chapter is an invitation to challenge every cultural belief you've absorbed about aging. It's not about reversing time — it's about resolving tension. Because when ripples realign, vitality returns.

And suddenly, "old age" begins to look a lot more like "old misalignment."

Aging Is Not a Clock — It's an Accumulation

For most of human history, aging has been measured in years. We've treated birthdays as milestones of decay — as if each tick of the clock automatically drags us closer to decline. The mainstream assumption is simple: The longer you live, the older you become. But this model is not just simplistic — it's fundamentally flawed.

Under the lens of Universal Energy Dynamics (UED), aging is not the result of time passing, but the consequences accumulating. It's not about the passage of years, but about the tension unresolved in your ripple field.

Let's clarify the paradigm shift:

- **Mainstream view:** Age is a number. It increases steadily and equally for all.
- **UED view:** Age is a reflection of ripple health. It increases only as much as unresolved interactions weigh down your field.

Every experience you have interacts with your personal ripple field. When an experience is processed, expressed, and integrated, its energy is released, and your field returns to coherence. But when an experience is suppressed, denied, or resisted, its energy lodges within you. It forms ripple knots — concentrated zones of field distortion that sap vitality, dull perception, and reduce coherence.

This buildup isn't visible on a calendar, but its effects are etched into your posture, your voice, your face, your thinking. It's why two people of the same chronological age can look, feel, and move like they're living decades apart.

Emotional Accumulation Is Biological Decay

Trauma, chronic stress, fear, guilt — these are not just psychological states. They are ripple states. And every time they loop unprocessed, they reinforce field drag, increase ripple friction, and accelerate cellular fatigue.

The body doesn't age because time tells it to. It ages because the field that sustains it becomes dense, distorted, and misaligned. Cells regenerate. Muscles repair. But when the instruction field — your ripple coherence — is burdened with unresolved tension, these systems misfire or slow down.

Analogy: A mirror doesn't become less reflective because years passed. It loses its clarity because dust settled — and nobody wiped it clean.

This is aging in the UED paradigm: You don't break down because time passed. You break down because the environment within you stopped clearing the residue of experience.

Time Is a Marker, Not a Mover

Yes, the Earth orbits the Sun. Yes, your cells follow circadian rhythms. But these periodic markers don't cause aging — they just coincide with it. The real cause lies beneath the rhythms — in your field.

When your ripple field is clean, responsive, and flexible, you feel younger — even if your biological years are many. When your ripple field is stagnant, heavy, or chaotic, you feel old — even if you're young.

In essence, aging is not what happens to you. It's what you carry.

This means aging is variable, reversible, and deeply personal — not inevitable. And it's not something to fear. It's something to understand.

You've been told your age in numbers. UED invites you to measure your age in tension. Because that's where real vitality is either lost or reclaimed.

Biological Decay and Energetic Misalignment

Aging, at its root, is not just about birthdays — it's about breakdowns in alignment. The wear and tear we associate with getting older is not simply due to "time" passing, but due to persistent misalignment in your energetic field. According to Universal Energy Dynamics (UED), biological decay is a ripple-level dysfunction — a distortion in the field that manifests in cells, tissues, and systems.

Let's unpack how this works.

Stress Isn't Just Mental — It's Cellular

Mainstream science has long observed that chronic stress accelerates aging. What's often overlooked is why. Stress isn't just a mental or emotional experience — it's a ripple disturbance that reverberates throughout your system. Every unresolved stressor tightens your internal field, creating ripple drag — a kind of friction that interferes with the seamless flow of energy.

This ripple tension doesn't float harmlessly in your aura — it pressurizes your biology. When you're chronically misaligned, your cells operate under energetic resistance. This creates inefficiency, impairs repair mechanisms, and slows regeneration. Over time, this builds into visible aging — not because of the years, but because of the ripples.

The Epigenetic Link — Genes React to the Field

Here's where UED and modern genetics unexpectedly align.

The field of epigenetics has shown us that gene expression is dynamic. It's not just your DNA that determines your fate — it's how your environment influences which genes are turned on or off. But what is this environment, really?

In UED terms, your ripple field is the first environment your biology responds to.

- Chronic emotional stress? That's ripple turbulence.
- Unprocessed grief or shame? That's ripple friction.
- Long-term dissatisfaction or misalignment with purpose? That's ripple drag.

These internal states signal your biology to switch into defensive or degenerative modes. Your genes respond not to "time" per se, but to the quality of field signals they're receiving.

So while medical science searches for anti-aging genes, UED points out the deeper truth:

The body doesn't age because of faulty genes. It ages because your ripple field teaches it to.

The Immune System: Your Inner Alignment Sensor

The immune system is often misunderstood as just a pathogen defense mechanism. But in UED logic, it functions more deeply as a coherence detector — a biological expression of your ripple field's clarity.

When your inner field is misaligned — due to unhealed trauma, prolonged fear, or toxic environments — your immune system senses this as instability. It shifts its behavior accordingly:

- Overactive immune states (e.g., inflammation, autoimmunity) arise when the field is chaotic.
- Underactive immune states (e.g., fatigue, susceptibility) occur when the field is depleted.

This means the health of your immune system isn't a function of your age — it's a function of your energetic integrity.

In essence:

The immune system listens not to your age, but to your alignment.

Ripple Drag and Metabolic Inefficiency

Why do people get slower, heavier, more fatigued with age?

The mainstream answer is "metabolic slowdown." The UED answer is ripple drag — the accumulated interference that makes energy processing harder.

Think of it this way:

- A clear ripple field = clean signal → efficient metabolism.
- A distorted field = noisy signal → erratic or slow metabolism.

You're not slowing down because you're old. You're slowing down because your system is working harder to maintain coherence against accumulated tension. Aging, then, is not decay by design — it's compensation overload.

The Two-Body Comparison

Let's bring this to life with a simple comparison.

Two individuals. Same chronological age. Same country. Same job.

- One has unresolved emotional tension, poor self-awareness, constant reactivity.
- The other practices ripple coherence — introspection, alignment, and intentional release.

The first walks with heaviness, poor immunity, and mental fog. The second walks with clarity, flexibility, and calm energy.

From a UED standpoint, this isn't mysticism — it's measurable.

- Blood markers of inflammation.
- Variability in gene expression.
- Strength of immune response.
- Brainwave coherence.
- Cellular regeneration rates.

These are all field effects.

So when people say, "He looks so young for his age," what they're intuitively sensing is ripple alignment — not skincare routines or miracle supplements.

Closing Bridge Thought

The idea that aging is inevitable is one of the most disempowering myths of modern life.

What if we saw aging not as a sentence — but as feedback? What if every wrinkle, every ache, every slowdown was simply a signal — a ripple asking to be acknowledged, resolved, or released?

Aging, then, becomes a conversation — not a countdown. And the true anti-aging strategy isn't to fight time — It's to restore alignment.

Memory Weight and Psychological Aging

When people talk about aging, the conversation often turns quickly toward memory loss, mental fog, and slowed thinking. But what if the real culprit isn't just neural degeneration, but ripple congestion?

From the lens of Universal Energy Dynamics (UED), aging of the mind is not purely biological or neurological. It's a field-level clutter—a consequence of unprocessed memory loops, unresolved emotional residues, and perceptual overload that congests the ripple bandwidth of awareness.

Slowness Is Not a Symptom of Age — It's a Symptom of Overload

Let's reframe the problem.

You don't necessarily think slower as you get older. Rather, you respond slower—because you're hauling more psychological weight through every interaction.

Every thought, every unfinished conversation, every emotional bruise you never tended to... these are open ripple loops. Like background apps running on a smartphone, they don't scream for attention—but they consume bandwidth and sap your system's responsiveness.

This isn't poetic metaphor. It's energetic mechanics.

Memory as Ripple Residue

In UED, memory isn't stored in your brain like files in a hard drive. It's encoded in your ripple field—as resonant imprints that linger unless resolved. The more unresolved impressions you accumulate, the heavier your field becomes.

- Positive memories can become distortions if they're clung to.
- Negative memories warp your now by looping fear or guilt.
- Even neutral memories can congest awareness if carried without relevance.

This is not to say memory is bad. It's to say that unintegrated memory creates psychological drag, which we interpret as "mental aging."

Emotional Loops Are the True Cognitive Burden

Why do some elders stay mentally sharp while others feel perpetually foggy?

It's not IQ. It's not even education. It's emotional integration.

Here's the UED insight:

Your brain is fast. Your field is what slows you down.

When your inner ripple field is saturated with open loops—regret, resentment, fears, worries—you lose clarity. You begin reacting to echoes rather than current signals.

- You forget not because memory failed—but because your awareness is diluted.
- You feel tired not because of age—but because you're reacting to unresolved layers.
- You hesitate, not from lack of knowledge, but due to perceptual interference.

Analogy: Too Many Open Tabs

Imagine a laptop with 57 browser tabs open, 12 apps running in the background, and several software updates pending. Even if the processor is brand-new, the machine runs slow.

Now apply that to your mind.

- Each unprocessed emotion = 1 open tab.
- Each suppressed memory = 1 draining app.
- Each fear of the future = 1 runaway script running in the background.

It's no surprise that you can't focus, think clearly, or make sharp decisions. You're not cognitively declining. You're ripple-degraded—like a clear lake clouded with debris.

Aging, in this case, is not biological deterioration. It's systemic clutter.

Field Congestion vs Brain Degeneration

Even mainstream science now accepts that neuroplasticity lasts a lifetime. Your brain can form new connections, adapt, and learn well into your 90s. So why do most people assume aging equals decline?

Because they conflate neural architecture with ripple architecture.

In UED terms:

- Neural activity is shaped by ripple alignment.
- Ripple misalignment creates chaotic signal inputs.
- Chronic misalignment = inefficient cognition.

Thus, the aging mind isn't just "weaker"—it's more disordered. Not because it lacks ability, but because it's buried beneath layers of unresolved information.

The Role of Silence, Reflection, and Release

You don't age because you forget.

You age because you refuse to release.

- You replay arguments from 10 years ago.
- You carry roles and identities long past their expiry.
- You burden your field with what-ifs and could-have-beens.

True rejuvenation starts not with brain games, but with field clarity.

This is why sages, monks, and artists often seem timeless: They regularly clear the cache.

Through silence, meditation, storytelling, movement, or even intentional solitude, they reset their ripple field. They don't cling to every memory. They don't react to every echo. They let it flow.

Psychological Youth Is Ripple Lightness

There's a kind of elder who walks like water — smooth, present, calm, sharp. And then there's the elder who walks like stone — heavy, reactive, stuck.

They might be the same age. Their cognitive tests might match. But their fields? Entirely different.

- One is light, open, flexible — low ripple drag.
- The other is dense, clenched, looped — high ripple residue.

So when someone says, "She's young at heart," they may be describing not emotion, but energetic agility.

Closing Bridge Thought

You don't need to keep every memory to stay wise. You need to integrate, not archive.

Aging is not forgetting. Aging is failure to let go of what no longer serves your present.

When you declutter your ripple field, your mind remembers how to breathe. And just like that, age lightens—not because you reversed time, but because you stopped accumulating it.

Youthfulness = Ripple Clarity

What if youth wasn't defined by the number of years since birth, but by how little you're carrying from those years?

In the lens of Universal Energy Dynamics (UED), youthfulness isn't about being young. It's about being clear. Not clear in thought alone, but clear in ripple field — free from backlog, free from unresolved loops, free from the drag of energetic residue.

We often say, "She's so full of life," or "He's young at heart." But what we're really pointing to is field coherence — a state of ripple alignment where the person interacts fluidly with the moment, without echoes of misalignment muddying the response.

Children Don't Carry Time — They Ride Ripples

Observe a child closely. They're not worried about what happened yesterday. They don't obsess over what might happen next week. They live entirely in ripple response — sensing, reacting, adjusting, flowing.

Children are not fast because they have more energy. They're fast because they have less drag.

- They laugh fully.
- They cry completely.
- They move on.

That's ripple clarity in action. And that's why their now is pure, their joy is immediate, and their fatigue is minimal.

Youth, in this sense, is not a stage. It's a state.

Age Accumulates When Resolution Stops

Every experience creates a ripple. But not every ripple gets resolved.

- You have a heartbreak — but never process the pain.

- You face failure — but keep replaying it in your head.
- You lose someone — but never integrate the absence.

Each of these becomes a stagnant wave in your field, reducing agility, slowing responsiveness, and aging your system.

So while years pass linearly, your ripple age grows based on what you haven't cleared.

Here's the insight:

Aging isn't a forward march — it's a weight pulling backward.

That's why some people at 30 already feel "old," and others at 70 still feel light, free, and ready to explore.

Youth Practices = Field Practices

The path to youthfulness isn't medical. It's energetic.

Certain practices don't make you young — they make your field clear.

- Meditation stills internal noise and resolves lingering loops.
- Joyful movement like dance, play, or nature walks re-aligns bodily ripples.
- Laughter and awe recalibrate your emotional field toward coherence.
- Deep sleep restores ripple balance by switching off conscious resistance.

These aren't anti-aging tricks. They're ripple cleansers.

Each time you laugh without condition, move without fear, or feel without clenching — you resolve a little more residue. You become lighter, faster, clearer. You feel young again — not because time reversed, but because drag reduced.

Ripple Resolution = Aging in Reverse

Let's flip the model.

You don't need to go backward to be youthful. You need to stop accumulating what makes you feel old.

In UED terms:

- Aging = unresolved ripple layers
- Youth = real-time ripple response
- Restoration = ripple resolution

So instead of fighting age, resolve misalignment:

- Finish conversations you've been avoiding.
- Forgive stories you keep retelling.
- Let go of roles that no longer serve your now.

Every act of energetic cleaning is an act of youthfulness.

Insight: Age Isn't Behind You — It's Within You

People think of age as a trail left behind. But in ripple logic, age is what you're still carrying — not what you've passed through.

That's why you can feel heavier after a single traumatic year than you did after ten joyful ones.

Your "age" is not a number. It's your drag coefficient.

- The more unresolved echoes you carry, the older you feel.
- The more aligned you are, the younger you become — regardless of birthdays.

The Real Secret of Eternal Youth

It was never about serums or supplements. It was always about flow.

Youth is the ability to move through experience without residue.

To feel fully, and then let go. To learn from pain, but not carry its imprint. To hold joy, but not cling to its memory. To align—again and again—with whatever ripple life sends next.

So when someone says, "You look younger these days," it may have less to do with skincare, and more to do with the fact that you've lightened your ripple load.

You are closer to your now. And the now — as you already know — has no age.

Can You De-Age Through Alignment?

Can you *actually* become younger — not metaphorically, but *measurably*?

From a Universal Energy Dynamics (UED) standpoint, the answer is provocative but clear:

Yes. But not by reversing time — by resolving tension.

In ripple terms, aging is not the passage of time, but the accumulation of unresolved misalignments. If that's true, then de-aging becomes possible not by chasing youth, but by re-aligning the ripple field — restoring the original coherence you were born with.

Let's reframe what it means to *grow old* — and more importantly, what it means to *grow light* again.

Regeneration Starts with Coherence

Every biological system thrives in coherence. Cells communicate better, metabolism functions optimally, and repair mechanisms activate when the ripple field is stable.

What disrupts this?

- Emotional clutter
- Mental loops
- Physical stagnation

- Constant external stimulation

These don't just age your mind — they age your biology.

But here's the opportunity: When the field clears, the body follows. When the mind settles, restoration begins.

Many people have experienced this:

- The cancer patient who changed their emotional outlook and saw remission.
- The 60-year-old who starts meditating and gains flexibility and energy.
- The person who walks away from a toxic environment and looks visibly younger within months.

These aren't miracles. They're ripple realignments. They represent the *reduction of inner resistance* — a de-aging at the energetic level that cascades down to cellular reality.

Real Examples: Alignment Over Intervention

Numerous documented cases reveal that:

- Telomere length (a key biological age marker) can increase through stress reduction, diet, and meditation.
- Skin health, hormonal balance, and even brain plasticity can improve well into old age.
- Sleep quality, often the first thing to go with age, rebounds quickly when emotional weight is lifted.

None of this requires time travel. It requires ripple harmony.

The body is not a victim of time. It's a mirror of your internal field condition.

Ripple Alignment as True Longevity Science

Longevity is not just about living long — it's about living clear.

What if aging was simply a lag in feedback? What if health problems were consequences still echoing from unresolved misalignments?

From UED's perspective:

- Aging = ripple drag
- Longevity = ripple clarity

The longer you stay in tune with your environment, your needs, and your truth, the less resistance you carry. The less resistance, the more agile your system — physically, mentally, emotionally.

That's not pseudoscience. That's energetic physics applied to life.

Practices That Re-Align and Rejuvenate

Let's make this practical. Here are ripple-alignment habits that directly contribute to de-aging:

1. Physical Movement (with Awareness)
    - Not just exercise — but *felt* movement: yoga, dance, walking with presence.
    - You're not just burning calories. You're releasing trapped ripples.
2. Emotional Detox
    - Journaling, therapy, expression, forgiveness.
    - Every unspoken truth is a ripple that won't dissolve until released.
3. Clean Perception
    - Seeing things as they are — not through filtered narratives.
    - Clarity reduces mental processing load, freeing energy for repair.
4. Stillness and Silence
    - Silence isn't absence — it's ripple recalibration time.
    - When the mind quiets, the field regains symmetry.

Each of these is anti-aging in the truest sense: They don't fix the surface — they resolve the origin.

The Real Anti-Aging Secret: Re-Alignment

We've been taught to think of age as irreversible decline.

But UED offers a different lens:

"You don't need to reverse time. You need to release what time left behind."

Forget creams and surgeries. Forget the obsession with youth as a look.

The real question is: How much unresolved ripple are you still dragging?

Because when that's cleared — when your system isn't echoing with stuck energy — you move like someone much younger. You feel fresh. You respond fast. You radiate lightness.

In a world chasing external restoration, the real key is internal realignment.

That's where youth lives. That's where age dissolves. That's where *you* return.

Closing Reflection

"You don't age by years — you age by echoes."

The calendar isn't aging you. The clock isn't cutting you down. What's truly wearing you out are unresolved echoes — ripple residues of emotion, tension, and dissonance that never got the chance to complete their loop.

Think of your body not as a machine with a fixed warranty, but as an instrument of energy. A ripple-sensitive system that was once in perfect harmony — clear, vibrant, alive. But as life happens, that clarity gets clouded. Tiny misalignments accumulate. They echo in your cells, in your posture, in your thoughts.

That's what we mistake for aging.

You didn't grow old — you grew layered. Layered with unspoken grief, unprocessed fear, unmet needs, and tension you had no time to release. And with each echo that stayed unresolved, your field lost some of its sharpness. The music dulled.

"Aging," in this view, isn't about passing through time — it's about accumulating unresolved feedback.

But here's the deeper truth: You are not the age your passport says. You are the clarity your ripple field holds. Which means — alignment can restore you. Lighten you. Even *de-age* you.

Let go of the myth that youth is behind you. What's behind you is just the trail of ripples you haven't cleared yet.

"Age is not about how far you've come — but how much you've left unresolved."

So start there. Resolve. Release. Realign. Not for vanity — but for vitality.

And ask yourself...

If aging is just unresolved ripple decay — Then what exactly is death? If nothing truly decays — only misaligns — What happens when the field lets go completely?

Part III — New Rules of Time

Chapter 12: "Don't Manage Time — Align It"

Self-Help Meets Physics — Productivity by Ripple Match

You were never running out of time. You were just out of alignment.

Productivity gurus told you to block hours, optimize routines, and conquer the ticking clock like it's your personal enemy. But have you noticed? Despite all the tools, hacks, and hustle — something still resists. You sit down to work, and your energy is scattered. You plan the perfect day, and life rearranges it anyway.

Here's the truth that most self-help books never told you: Time doesn't obey your calendar. It obeys your field.

Under the lens of Universal Energy Dynamics (UED), time is not a stream you manage — it's a consequence of ripple alignment. Every action you take either matches the readiness of your field or fights it. When it matches, things flow. When it doesn't, things break, delay, or exhaust you.

In this chapter, we flip the script on productivity.

You'll discover that what looks like procrastination may just be field resistance. That peak performance isn't about pushing through — it's about knowing when your ripple is ready. That success isn't built on discipline alone, but on sensing when the energy says: now.

Forget time management. Welcome to ripple match.

Let's realign the way you work — not by chasing minutes, but by removing the inner drag that makes every minute feel like a marathon.

The Myth of Time Management

You Don't Control Time — You Respond to It

For decades, we've been told the secret to a better life is "time management." From day planners to productivity apps, we're taught to budget hours like money, schedule everything into neat blocks, and squeeze every drop out of the 24-hour cycle. Wake up early. Plan your priorities. Eliminate distractions. Optimize routines.

But here's the uncomfortable truth:

You don't actually manage time. You manage your field's ability to respond.

Let's unpack this. Time, as explained by Universal Energy Dynamics (UED), is not a container you move through — it is a field delay that arises from ripple resistance. When your attention, energy, and intent are in misalignment with your surrounding ripple field, everything takes longer. You're not running out of time — you're pushing against it.

So, what does that mean for traditional time management techniques?

Take the popular practice of "time blocking." On paper, you assign a task to a fixed slot — say, writing from 9:00 to 11:00 AM. But at 9:03, you're stuck. Nothing flows. Your body is restless, your mind is scattered, your emotional field resists. You blame lack of discipline. But what if discipline isn't the problem? What if your field isn't ready?

UED teaches that ripple alignment is everything. When your internal state (intent, energy, emotional readiness) matches your external conditions (environment, task demand, field rhythm), action feels natural. Smooth. Effortless. That's when you're "in the zone." Time bends — or disappears.

But when the match is off — when you force an action outside of your readiness — you experience friction. The mind drags. The task resists. Minutes feel heavy. Hours are wasted.

You didn't fail at time management. You failed at ripple alignment.

Let's use a simple analogy. Imagine a river. The current flows in a direction with its own rhythm and pace. You're in a boat with a paddle. Time management says: "Move at 9 knots, whether the river agrees or not." Ripple alignment says: "Feel the current. Position yourself to flow with it. Paddle when the field supports you."

Which one is more intelligent?

Self-help traditions focus on controlling behavior. UED focuses on reading the field. The difference is profound. Traditional productivity tries to overpower nature's rhythm. Alignment-based productivity listens to it — then acts with precision.

This doesn't mean you sit idle, waiting for a cosmic signal. It means you develop the skill of sensing your field — the subtleties of readiness, friction, and flow. You learn to act when action is supported — and rest or realign when it's not.

Let's make it practical:

- Struggling with writing? Your field may need clarity — not force.
- Exhausted after 15 minutes of focus? You're misaligned, not weak.
- Can't finish a task you blocked time for? The ripple may be unfinished elsewhere.

True productivity doesn't come from chasing the clock. It comes from aligning with the ripple that's ready to move.

So, the next time you look at your to-do list, ask not "Do I have time?" Ask: "Am I aligned?"

Because in the UED framework, time doesn't give permission. Alignment does.

Productivity as Ripple Matching

Flow Happens When You're Field-Ready

Productivity is often misunderstood as doing more in less time. But in the Universal Energy Dynamics (UED) framework, productivity isn't about pace — it's about precision. It's not about "how fast" you do something, but how well your inner state matches the ripple field of the task.

This is ripple matching. And it's the real secret behind flow, creativity, and high performance.

The Myth of the Grind

Let's challenge a modern obsession: hustle culture.

You've been taught to show up at 5:00 AM, power through fatigue, and grind your way to success. But how often do you finish those grueling hours feeling truly fulfilled, energized, or proud of the output? How many times have you sat for hours, forcing yourself to write, code, build, or brainstorm — and still gotten nowhere?

That's ripple resistance.

When your field — your emotional state, attention, physiology, and subconscious — isn't in resonance with the task at hand, even simple work feels like climbing uphill. That's because the ripples inside you (intent, readiness, energy) are out of phase with the external demand (task, environment, expectation).

In this state, you're not unproductive because you're lazy. You're unproductive because your ripples aren't aligned.

Now contrast that with those rare but magical moments when you enter a flow state.

You sit down to write, and words pour out. You begin a conversation, and insight unfolds. You tackle a problem, and solutions arrive almost faster than you can act on them.

What changed? Not the clock. Not the calendar. The field did — and you aligned with it.

The Ripple-Matching Principle

UED introduces a radical but practical idea:

"When task intent and field readiness resonate, time collapses and productivity expands."

In simple terms — when your inner readiness matches the ripple state of the task, you don't need more time. You need less. Because the work flows.

Think about it like surfing. You don't create the wave. You don't control the ocean. You learn to sense the right moment — the perfect alignment — and then you ride it. That's ripple-matching in action.

Forcing a task when your field is closed is like trying to surf on still water.

Creative Work as Proof

Let's take writing as a case study.

Writer's block is not a lack of discipline — it's a field mismatch. Your cognitive ripples are not yet aligned with the theme, the emotion, or the message you're trying to express. The more you force it, the more resistance you feel. You might type words, but they feel hollow. You edit endlessly, never satisfied. Hours pass — progress doesn't.

Now compare that with a moment of creative burst. A single hour may yield paragraphs of rich, clear, coherent thought. You forget time. You forget effort. It just moves.

What happened?

The ripple field inside you resonated with the task's ripple pattern — and alignment replaced effort.

Peak Performers Know This

Whether consciously or not, top athletes, artists, leaders, and innovators all ride the rhythm of ripple matching. They don't act just because the clock says so — they act when their field and the field of action align.

- A footballer senses when to strike — not by clock, but by field awareness.
- A CEO chooses the right moment for a pitch — not by schedule, but by sensing readiness.
- A dancer performs not by counting minutes, but by moving with inner timing.

They may have routines, yes. But their real mastery lies in feeling the ripple readiness — and acting at the peak of resonance.

The Cost of Misalignment

Now let's look at the other side — the cost of ignoring ripple mismatch.

Trying to execute misaligned tasks results in:

- Mental fatigue: constant inner friction.
- Low quality output: your field can't hold clarity.
- Time dilation: tasks take longer and feel heavier.
- Emotional drain: frustration accumulates, and confidence drops.

This is not failure. It's ripple turbulence. You're not doing wrong — you're doing out of rhythm.

From Time Slots to Ripple Slots

So how do we apply this?

Instead of strict to-do lists, begin to build Ripple Readiness Maps:

- What tasks match your peak alignment times?
- What environments enhance your field clarity?
- When does your emotional and cognitive state support deep focus vs ideation?

Design your day not by hours, but by alignment windows.

Instead of saying "I'll write from 9 to 11," try: "I'll write when I feel aligned and clear — and rest or prepare until then."

That shift alone can save you hours.

Final Insight

"Flow is not a mood — it's a field state."

True productivity emerges not when you push harder, but when you align deeper. It's not about doing more — it's about doing right when it's right.

That's the new game. Not managing time — but mastering resonance.

Calendar ≠ Readiness

Why Deadlines Often Fail

"Just because it's on the calendar doesn't mean it's ready to happen."

Let's confront one of the most widespread delusions in the world of productivity: We assume that scheduling something makes it real.

You slot an important task for Tuesday 2 PM. You set goals for the month, quarterly deliverables, five-year plans. It's all in the calendar — colored, synced, shared, systematized.

But then 2 PM arrives. And you're not ready.

Your body is anxious. Your mind is distracted. Your emotion resists. Your environment stutters.

And still — you try to push through. Because the calendar said so.

This is where most productivity collapses:

We mistake scheduled time for ripple readiness.

Deadlines Don't Create Readiness

Let's break this down using UED.

Every action — whether it's writing a chapter, launching a business, having a hard conversation, or simply showing up with presence — requires ripple alignment. Your inner field (attention, energy, emotion, clarity) must stabilize into resonance with the action field (task, environment, intent).

Only then does progress feel possible — and real.

But a calendar doesn't know your field.

It knows only repetition and allocation. It's a loop. Not a sensor. Not a permission slip.

You may schedule transformation — but unless your ripple field agrees, all you'll feel is friction.

Why Deadlines Often Fail

How many times have you "planned" to finish something, only to reschedule, delay, or force it through poor quality?

Deadlines often fail not because of lack of will — but because the field wasn't aligned. Mentally, you knew what to do. But emotionally, energetically, situationally — you weren't there.

UED reveals this truth:

"Tasks don't happen on time. They happen in alignment."

You can't force a seed to sprout just because the calendar says it's Day 5. It sprouts when the conditions match its internal readiness. Sun. Soil. Water. Stillness.

So it is with your actions.

Field Pressure vs Mental Pressure

We've been conditioned to interpret mental pressure as discipline.

- "Just do it anyway."
- "Push through the block."

- "Discomfort is growth."

But these are dangerous generalizations when divorced from field awareness.

Sometimes, discomfort is a sign of fear worth overcoming. But often, it's a signal: Your field is unstable. Your internal ripples haven't reached alignment yet. Acting now will produce distortion, not evolution.

UED teaches us to distinguish between:

- Field Permission: A sense of resonance, clarity, flow — even amidst challenge.
- Mental Pressure: A self-imposed "should," rooted in expectation, fear, or rigidity.

They feel different. You know when you're in flow. And you know when you're fighting it.

Permission ≠ Procrastination

Now let's clarify something crucial.

This is not a license for endless delay or spiritualized avoidance. This isn't saying: "I'll wait for the stars to align" or "I wasn't feeling it."

That's not alignment — that's avoidance disguised as intuition.

True field readiness comes from active awareness and preparation:

- You sit with the resistance.
- You stabilize your ripples (through silence, breath, movement, or reflection).
- You watch for the click — the soft yes from the field.

And when it comes — you move. Quick. Clean. Focused. The work flows, not because it's easy — but because it's ripe.

The Calendar Was for Cattle

Fun etymology twist: The word *calendar* originates from the Latin *calendae*, which referred to the first day of the month — when debts were due.

Calendars were tools of accounting and control — not awareness or alignment.

Today's obsession with tightly planned hours, Gantt charts, and micro-managed routines comes from the industrial era, not from the reality of human ripple fields.

Machines run on fixed inputs and outputs. Humans don't.

We are field-sensitive, context-aware, emotionally layered ripple systems.

"You're not a clock. You're a tide."

Practical UED Insight: Create Alignment Blocks, Not Time Blocks

Instead of scheduling by arbitrary hours, experiment with alignment blocks:

- Begin your day with ripple check-in: What's my state? What wants to emerge?
- Block time only when inner readiness feels probable — not just when you're free.
- Design your schedule around resonance zones — periods where certain types of tasks naturally align with your energy.
- Leave white space — not as filler, but as buffer for ripple recalibration.

This isn't laziness. It's efficient alignment.

Final Takeaway

The calendar is useful — but only when it reflects your field. If it dominates your field, it becomes a source of misalignment, burnout, and self-blame.

Remember:

"A scheduled hour means nothing if your ripples aren't ready."

What gets done is never what was *planned* — but what was aligned.

Alignment Hacks for Real Efficiency

Train Your Field — Not Just Your Schedule

"Don't force the task. Feel the field."

Traditional self-help treats you like a machine that just needs better scheduling.

- Wake up at 5 AM.
- Write your to-do list.
- Crush your goals.
- Track your habits.
- Reward your discipline.

But here's the truth through the UED lens:

You're not inefficient because of bad time management — you're inefficient because your field is misaligned.

You're trying to execute before your ripples are ready. You're running on timers, not resonance.

And this is exactly where alignment hacks come in.

Let's rewrite efficiency not through control — but through coherence.

Replace Rigid Routines with Ripple-Sensing Rituals

Routines can be grounding. But rigidity kills alignment.

- A routine says: *Do it every day at 7 AM no matter what.*
- A ritual says: *Tune into your field, and activate when it clicks.*

What works today may misfire tomorrow if your inner state has shifted.

Instead of forcing yourself into routines like a discipline robot, learn to listen to the field.

The most successful creators, athletes, and leaders often follow rituals of readiness — not robotic repetition.

Practice 1: Field-Check Journaling Before Planning

Before planning your day, do this:

1. Sit in silence for 2 minutes.
2. Ask: *Where is my energy right now?*
3. Write down:
    - What feels light?
    - What feels heavy?
    - What feels incomplete?
    - What's calling me?

This ripple journaling becomes a field map. Instead of planning based on urgency or guilt, you're planning based on alignment readiness.

Try it for a week — you'll be shocked how your output improves, even if your schedule looks looser.

"You're not writing your to-do list — you're tuning your antenna."

Practice 2: Micro-Realignments (Real-Time Ripple Tweaks)

Throughout the day, you drift. You get pulled out of alignment by:

- Messages
- Emotional noise
- Cognitive loops
- Sudden demands

Traditional advice says: *Push through it.* UED says: *Pause. Realign. Proceed.*

Micro-realignments are short, intentional acts to re-stabilize your field:

- 90-second breath reset (4-4-4-4 box breathing)
- Walk outside for 3 minutes
- Sit in total silence with eyes closed
- Shift posture and release tension
- Drink water consciously
- Gaze at a plant or sky — tune back to nature's rhythm

These aren't productivity hacks — they're field hygiene.

You're not losing time by pausing. You're reclaiming coherence.

Practice 3: The "Wait for Click" Method

You've felt it before:

- You stare at the task. Nothing moves.
- You try to push. Still blank.
- Then suddenly — something *clicks*. The inner resistance drops. Energy flows.

That's the field alignment point.

The "Wait for Click" method trains you to:

1. Delay action until that micro-alignment is felt.
2. Use the waiting time for low-stakes ripple resets.
3. Let the work emerge *when it's ripe* — not when the clock screams.

It's not procrastination. It's precision.

This method builds deep trust in your ripple intelligence.

Practice 4: Anchor Tasks to Energy, Not Time

Rather than saying,

"I'll write at 2 PM," try saying, *"I'll write when my ripple feels open and quiet — usually after lunch + walk."*

Identify when your creative, strategic, or emotional fields are most stable — and anchor tasks there.

Build your day like a field garden, not a task grid.

Examples:

- Mornings: High clarity? Anchor creative work.
- Midday: Energy dip? Anchor admin or walk.
- Evening: Insightful? Anchor reflection or soft creation.

This replaces "discipline struggle" with field orchestration.

Why Ripple-Aligned Habits Outperform Willpower

Willpower is a limited battery. Alignment is a renewable current.

Habits that are forced by discipline often collapse under pressure.

But ripple-aligned habits — like movement that feels joyful, writing that rides a wave, or deep work that begins with silence — they stick. They become natural expressions, not enforced behaviors.

"The best habits aren't pushed into place — they unfold from the field."

That's why someone meditating at 4 AM isn't showing off discipline — they're showing field alignment that made it feel easy.

Final Insight: Efficiency = Field Trust

The greatest shift you can make is this:

- From "How do I get more done?"
- To "How do I align with what's ready to be done?"

Train your field, not your clock. The more you refine ripple awareness, the more you'll:

- Start fast (when it's time)

- Move light (without friction)
- Finish strong (without burnout)

The world teaches efficiency as a fight against time. UED teaches it as a dance with resonance.

Burnout = Prolonged Misalignment

Why Time Blocks Don't Save You

"You're not tired because you did too much. You're tired because you forced too much misaligned doing."

In today's hustle culture, we're taught to organize our day into neat little time blocks. Productivity apps, planners, Pomodoro techniques — all promise to guard you from burnout.

But despite these systems, burnout is more rampant than ever.

Why?

Because burnout isn't about hours. It's about misalignment.

Burnout = Ripple Collapse, Not Calendar Overload

In the Universal Energy Dynamics (UED) framework, every action we take is a ripple response. Ideally, it arises from coherence between:

- Inner readiness
- Field resonance
- Intent clarity

When these are misaligned — especially over long durations — the ripple field becomes unstable. That instability doesn't just slow you down. It tears you apart, subtly at first, then all at once.

You're not "running out of time" — you're exhausting your ripple tolerance.

Burnout = prolonged violation of your ATT (Alignment Threshold Time)

If you repeatedly act before your field is ready, you build resistance. This ripple friction accumulates like invisible stress debris.

Time blocks, deadlines, and external schedules ignore this inner readiness. They force scheduled effort on top of ripple lag — and call it discipline.

But discipline without alignment is violence.

Why Time Blocks Fail

Scheduling assumes uniform readiness. But human fields are dynamic, emotional, and ever-shifting.

- You may schedule "Deep Work from 9–11 AM." But what if your ripple is noisy at 9 AM? What if you're carrying unresolved tension from yesterday?

Time blocks don't account for:

- Emotional residue
- Mental clutter
- Environmental noise
- Physiological fatigue
- Unacknowledged misalignments

So, you sit down to work, but nothing moves. You push harder. You get something done — but at the cost of ripple harmony. Repeat this over days and weeks, and you burn out not from lack of time, but from disregard of your energetic bandwidth.

Burnout Symptoms Through the UED Lens

Let's reframe common burnout signs:

- Mental fatigue = ripple congestion from unresolved thought loops
- Emotional dullness = loss of resonance sensitivity
- Physical depletion = metabolic drag from misaligned tasks
- Creative block = disrupted signal clarity in ripple field

- Cynicism or detachment = protective dissociation from ripple dissonance

You aren't just tired — you're field-fractured.

And no amount of rescheduling will fix that.

Real Recovery = Time Off + Realignment

Rest is not just the absence of activity. It is the presence of repair.

That means:

- Emotional alignment
- Mental stillness
- Physical coherence
- Ripple reset

People who "take time off" but stay mentally stressed don't recover. They stop *working*, but they don't stop *resisting*.

Recovery = space where your ripple field is allowed to re-synchronize with itself

Rest is not idle. It's sacred.

Just like the ocean needs stillness to settle, your inner field needs undisturbed pause to regain resonance.

Field-Aware Recovery Practices

1. **Active Rest:** Activities that bring ripple harmony — walking in nature, playing music, gardening, cooking without pressure.
2. **Ripple Detox:** Journaling emotional blocks, expressing unspoken thoughts, creating safe silence for recalibration.
3. **Field Re-sensing:** Breathwork, mindfulness, solitude — allowing attention to re-enter the ripple pattern calmly.
4. **Energy-First Planning:** Begin your day by sensing energy — not

scanning calendars.

You don't heal from burnout with naps and Netflix. You heal by re-entering alignment.

Final Insight: Efficiency Without Alignment = Erosion

What you gain in speed during misaligned productivity, you lose in long-term coherence.

You may get the task done. You may hit the target. But you pay with ripple debt — and eventually, you crash.

True efficiency isn't fast work. It's harmonic work. Work that doesn't exhaust your field — it extends your flow.

So next time you feel burned out, don't ask:

"How much time do I need off?"

Ask:

"Where did I lose alignment?"

Closing Reflection

"You don't need more hours. You need fewer obstacles."

Time was never your enemy. Misalignment was.

For too long, we've been sold the fantasy that if we just managed our time better, we'd win the day. More hours. Better planners. Tighter schedules. But what if that's not the issue at all?

You don't need more hours. You need fewer obstacles to your field's alignment.

In the UED lens, time is not a real commodity. It's a byproduct of ripple delay — of how misaligned your internal state is with your intended action. When your ripple field is clear, your work flows effortlessly. When it's blocked, even a five-minute task feels impossible.

You are not fighting the clock — You are navigating your resonance.

This chapter wasn't about doing less or abandoning responsibility. It was about reorienting your relationship with time. Stop managing it like a resource. Start sensing it as feedback.

- If things are stalling, it's not poor planning — it's poor ripple match.
- If your energy crashes midday, it's not a lack of time — it's a misfit between your field and your focus.
- If you're burning out, it's not your calendar — it's your friction.

Let go of the illusion that productivity lives inside the ticking of a clock. Real output — real momentum — lives in alignment.

Final insight:

"When you stop chasing minutes and start sensing alignment, you don't just save time — you dissolve it."

Time becomes irrelevant when your field is tuned. Because the real game isn't managing your hours. It's learning to move when your being says yes — and pause when it says no.

Not lazy. Not indulgent. Just energetically precise.

Chapter 13: "Time Doesn't Heal. Alignment Does."

Rethink Trauma, Pain, Growth — Healing is Ripple Reset, Not Ticking

"Time never healed anyone. It just blurred the noise — or buried the wound."

We've all heard it. That reassuring phrase offered like a bandage: *"Give it time."* As if minutes and months were medicine. As if healing were a passive process that arrives simply because the clock kept moving. But here's the uncomfortable truth — time doesn't heal. It just numbs the sharpness, muffles the pain, hides the ripple beneath the noise of daily routine. It gives illusion of distance, not resolution.

In the framework of Universal Energy Dynamics (UED), healing isn't about waiting — it's about *realigning.* Every trauma, heartbreak, or breakdown leaves a ripple misalignment — a disturbance in your inner field. Time alone does nothing to reset it. You carry that echo, that friction, until it's consciously addressed, emotionally processed, and energetically integrated.

Imagine a musical instrument that's out of tune. Does it retune itself just because a day passed? Or does it stay out of sync until someone realigns the strings? Likewise, your inner field doesn't restore with time — it restores with *clarity*, *permission*, and *presence.*

The myth of "time healing" is a comforting story — but one that keeps people stuck. We wait. We distract. We endure. All while the misaligned ripple loops inside, subtly shaping our behaviors, reactions, and sense of self. The result? Years pass. The wound may stop bleeding — but it still hasn't closed.

This chapter isn't about recovery timelines. It's about healing as ripple reset. Whether you're dealing with emotional trauma, chronic stress, or just the weight of unresolved grief, the real path forward isn't time-based — it's field-based. When the field realigns, pain fades. When resonance returns, so does peace.

So before you check the calendar again, pause. The moment doesn't need to pass — it needs to process. Let's explore how healing really works when you stop counting days and start sensing alignment.

What We Mistake as Healing

Scars Aren't Always Signs of Healing — Sometimes They're Evidence of Avoidance

There's a cultural narrative we've all inherited — a silent prescription for pain: *"Let time do its work."* We assume the ticking of clocks and the passing of days will slowly untangle the mess inside. But what if time is simply putting space between you and the wound — not actually closing it?

The truth is sobering: time creates distance, not resolution.

Just because something fades doesn't mean it healed. Just because you can function again doesn't mean you've restored your internal field. In the Universal Energy Dynamics (UED) framework, healing is not about duration — it's about ripple restoration. When we suppress pain, deny it, or distract ourselves from it, the unresolved ripple doesn't vanish. It freezes. It loops silently beneath our awareness, showing up in other places — tension in the body, defensive reactions, restlessness, even sudden breakdowns over minor triggers.

We call this *coping*. But coping without alignment is simply field evasion.

Let's look at a scar. It's often treated as proof of healing — the wound is closed, the body has moved on. But what if the scar is simply a physical reminder of what was never fully integrated? Sometimes, that scar is a hardened echo — not resolution, but residue. A skin-level ceasefire masking a ripple war beneath.

Pain that isn't processed is pain that keeps broadcasting — subtly warping your perception, guiding your decisions, and shaping your relationships. You stop trusting. You second-guess joy. You prepare for things to fall apart again — not because you're paranoid, but because your field still remembers. The event is in the past; the ripple is not.

One of the greatest myths is that *feeling better = being better*. You might feel lighter today because you're distracted, busy, or numb. But if the emotional loop wasn't closed — if the ripple wasn't resolved — you're still carrying the field distortion. The absence of pain isn't healing. Sometimes, it's just avoidance with better lighting.

"Give it time" is a well-meaning phrase, but dangerously incomplete. What we need to give pain is not *time*, but space to realign — time alone delays, but alignment transforms.

Here's an analogy: imagine your favorite song is playing on a loop in another room, but you mute it. You can no longer hear it, but the song hasn't stopped. It's still going — shaping the space, defining the atmosphere. That's what unresolved trauma does. Time mutes the volume. Alignment is what stops the song and changes the playlist.

It's not enough to wait. Waiting only shifts the sensation — not the signal. Healing demands interaction. Reflection. Processing. Emotional engagement. Field recalibration.

So, if you've ever wondered why something that happened "so long ago" still affects you... it's not because you're weak, broken, or overly sensitive. It's because time didn't fix it. And it never could.

In the ripple framework of UED, real healing begins not with counting the days, but with confronting the distortion. You heal not when time passes — but when you pass through the ripple that still holds you.

Ripple Repair — The Real Path to Healing

You Heal When You Realign — Not When You Wait

Healing, in its deepest form, is not something that happens *to* you — it's something that happens *through* you. Contrary to the passive myths of "time as a healer," true recovery is an act of energetic participation. In the UED framework, healing is not linear, logical, or calendar-bound. It's a ripple realignment process, where the distortions caused by pain, loss, shock, or betrayal are gradually or suddenly brought back into coherence.

Let's begin with a definition: Healing = restoring ripple clarity. Your personal field — mental, emotional, biological — is constantly resonating with the environment, people, thoughts, and actions around you. When aligned, the field is fluid, adaptable, and responsive. You feel "in tune" — calm but alert, sensitive but not overwhelmed. But when something disruptive enters — heartbreak, failure, trauma — your field fractures. Certain parts tighten, others scatter. You begin reacting instead of resonating.

This is misalignment. And misalignment isn't invisible. It shows up as symptoms:

- Illness that doesn't trace back to clear biological causes
- Depression that feels like a fog without edges
- Anger loops that hijack your reactions
- Emotional numbness that freezes your ability to connect

Each of these is a field response, not just a psychological or medical event. They are ripple turbulence echoing from unresolved interaction.

Now, here's the part mainstream healing misses: You don't heal by avoiding the ripple. You heal by re-entering it — consciously, compassionately, and courageously.

This is where true healing practices come in. Not as magical fixes, but as tools for ripple reset. Breathwork, for instance, stabilizes your internal rhythm — slowing down inner turbulence and giving your field a chance to harmonize again. Movement (yoga, dance, walking) restores kinetic coherence — allowing stuck emotions to move physically. Emotional expression — whether through tears, art, or words — is field ventilation. It breaks the loop of internal pressure and invites energy to flow again.

And sometimes, healing arrives as a moment of insight. A sudden clarity. A conversation. A line in a book that pierces through your noise and shifts your perspective. That moment doesn't take time — it takes *alignment.* This is instant ripple recalibration — a cognitive snap that reorders your inner field.

Other times, healing is slower. Gradual reintegration. Like tuning an instrument string by string. You test. You wobble. You stretch. But each act — journaling, therapy, solitude, mindful connection — nudges your field closer to harmony. This is *conscious ripple work.*

And here's the most misunderstood part: You can't fake healing. You can suppress symptoms. You can intellectualize pain. You can even distract yourself with productivity or positivity. But if your field hasn't realigned, the echo remains. Unspoken. Unprocessed. Waiting for the next trigger.

Think of it like noise-canceling headphones. The disruption is still out there — you've just muted your awareness of it. But healing isn't silence through suppression — it's silence through resolution. The ripple doesn't disappear because you drowned it — it settles because you engaged with it until it integrated.

Let's make this practical.

What helps ripple repair?

- Stillness — allows you to hear the inner ripple clearly.
- Expression — gives the echo an exit.
- Witnessing — being seen or heard by another brings external coherence to your field.

- Forgiveness — not for others, but to release the bind that holds your ripple hostage.
- Presence — when you're fully here, ripple distortion has less space to loop.

These aren't spiritual clichés. They're field interventions.

And just as pain can fragment you instantly, coherence can restore you instantly — or steadily. A sudden breath of truth. A deep sob. A reunion. A realization. Each one realigns the field, bringing the echo to rest.

The myth we must now bury is this: That time, on its own, fixes what broke. It never did. What *really* heals you is your willingness to return to resonance.

In the end, you're not waiting for the years to pass. You're waiting for your ripples to harmonize again — for the noise inside to quiet, not because it's forgotten, but because it's finally heard, felt, understood, and released.

That's healing.

Trauma Is a Time Loop

Unresolved Pain = Ripple Echoes Without Closure

Trauma is often misunderstood as a memory of a past event — something you "went through." But in the UED framework, trauma is not the past. It's the present ripple distortion caused by an unresolved event. You're not living in memory — you're living in misalignment that refuses to fade because it was never brought into coherence.

Let's break this open.

Imagine sound waves bouncing in a cave. You shout once, but the echo repeats — bouncing off the walls long after the source is silent. That's trauma. Not the original scream, but the unending reverb of it.

In ripple terms, trauma is a feedback loop your field never absorbed or released. The initial event — loss, violence, betrayal, fear — creates a shockwave in your system. If, at that moment, the field was not supported (emotionally, relationally, cognitively), the ripple couldn't resolve. So it froze in misalignment, and began looping.

This is why trauma feels timeless. You're triggered by a word, a tone, a gesture — and suddenly, you're back there. Not logically, but energetically. Your system isn't "remembering" — it's reliving. The ripple pattern hasn't changed, so the feedback keeps repeating.

In UED terms, this is a stuck ATT — Alignment Threshold Time. Just as physical ATT measures the delay in perception, emotional ATT is the delay in integration. The field encountered a spike — but never realigned. So the event remains "live," like an open tab on your browser, draining energy in the background.

Let's go further.

You're not haunted by the past. You're haunted by unresolved present feedback shaped by the past. It's not that you *can't let go* — it's that your field *won't* let go until the ripple is heard, acknowledged, and absorbed.

This explains why trauma shows up in loops:

- You meet similar people.
- You fall into familiar conflicts.
- You react in disproportionate ways.

These aren't random. They're ripple recall mechanisms. The field is trying to complete what was left incomplete. Every trigger is a chance for re-alignment — not just a trap.

Analogy: Think of a song with a skipped beat. Your body keeps waiting for the drop that never comes. That skipped beat becomes a loop, repeating over and over, trying to land, to resolve — but it never does. Until you stop, listen, absorb — and recompose.

From the outside, trauma looks like overreaction. But from the inside, it feels like no reaction is ever enough to end the loop. That's because your field doesn't need more reaction — it needs realignment.

This is where conventional time fails us. You might go ten years without an episode, and then — suddenly — you're overwhelmed again. Why? Because years don't resolve ripples. Alignment does. The wound was never healed — it was only buried. And the field remembers.

Here's the powerful realization:

You are not trapped in the past — the past is trapped in you.

And it's not metaphorical. It's physical, emotional, neurological. Trauma etches itself into the field — into posture, breath, tone, instinct. It's stored in how your body tightens, how your voice changes, how your awareness narrows under threat.

But here's the light: You can break the loop.

How?

- **Safety** — A ripple can't resolve in chaos. You need calmness, even momentary, for realignment to begin.
- **Witnessing** — When someone hears you, sees you, stays with your ripple, your field co-regulates.
- **Expression** — Naming the echo releases its grip. Silence hardens the loop — expression softens it.
- **Movement** — Physical release interrupts the feedback. You break the echo with a new resonance.

And above all: Presence. Not dissociation. Not distraction. Not suppression. Presence.

The moment you fully attend to the ripple, the loop starts to dissolve. It's not easy — but it's possible.

You're not broken. Your field is just stuck in replay — waiting for you to sing a new note.

Let's leave with this truth:

Healing from trauma isn't moving on — it's moving back *into* the ripple with awareness, and finally letting it resolve.

Time didn't fail you. It was never the healer to begin with. You are.

Growth = Ripple Reintegration

Real Growth Isn't Getting Over It — It's Becoming Coherent With It

In the aftermath of pain, the world often tells you to "move on." Let it go. Get over it. But true growth doesn't work that way — you don't grow by erasing the ripple, but by reintegrating it. In UED terms, growth is ripple coherence restored after rupture.

Let's begin here:

Real growth isn't forgetting — it's reframing.

It's not distancing from what hurt you — it's becoming whole around it.

Pain is not the problem. Pain *resisted* is. When ripple misalignment is denied, suppressed, or bypassed, it festers — repeating in emotional loops, toxic behaviors, and unresolved beliefs. But when acknowledged and reintegrated, that same ripple becomes a new pattern of resonance — a deeper understanding, a recalibrated self.

Think of it like music. A discordant note can ruin a melody — unless it's resolved by the next chord. Suddenly, what once jarred becomes the very tension that made the harmony beautiful. This is what ripple reintegration does. It doesn't remove the rupture. It completes it.

Let's go deeper.

Pain, When Aligned, Becomes Wisdom

Emotional pain is the echo of a misfit — your field encountered something it couldn't absorb. Growth is the process of going back, attending to that echo, and bringing it into context, meaning, and resolution. This is how wisdom is born.

- A betrayal, when reintegrated, becomes the seed of discernment.
- A loss, when faced, becomes the doorway to depth.
- A failure, fully absorbed, becomes the foundation of resilience.

UED insight: Every unresolved ripple carries energy. That energy can distort you — or, if re-patterned, empower you. Growth is simply this: choosing to realign instead of recoil.

Reintegration ≠ Resistance

It's tempting to resist what hurt us. But resistance keeps the ripple alive — it sharpens the edge of memory, hardens our identity around the wound, and splits the self into "before" and "after."

Reintegration, by contrast, softens the field. It welcomes the echo, listens to it, and lets it settle into the whole. You don't have to "be over it." You just have to be with it — differently.

That's the turning point: When the ripple is no longer an intruder, but a participant in your coherence.

Example: A person who was silenced in childhood may grow up fearful of expression. But if they revisit that echo with awareness — not as a victim, but as a field seeking alignment — they begin to speak. Not just to others, but to themselves. That voice becomes their growth. Not in spite of pain, but *because* of it.

Post-Traumatic Growth vs. Unresolved Adaptation

In psychology, we speak of *post-traumatic growth* — where individuals, after deep suffering, report increased strength, purpose, and empathy. But not everyone reaches this. Many remain in *unresolved adaptation* — coping, compensating, surviving.

What's the difference?

- Unresolved adaptation is a life designed *around* the ripple — avoiding triggers, staying guarded, losing joy.
- Reintegration is a life designed *with* the ripple — incorporating the scar into the art, the wound into the wisdom.

UED explains this as a shift from ripple avoidance to ripple inclusion. The field no longer fights the echo — it re-patterns around it. The dissonance becomes a new chord.

Growth, then, is not becoming someone "better." It's becoming someone more complete. Not newer — but truer.

You Grow in How You Carry the Ripple

The experience may not change. But your relationship to it does. You go from:

- *Why did this happen to me?* → *What did this reveal in me?*
- *How do I make it go away?* → *How do I make space for it within?*

That's the architecture of growth: Not walls to keep the ripple out, but rooms to let it live safely inside.

This doesn't mean you celebrate pain. But you consecrate it — give it meaning, a place, a purpose.

Because here's the truth:

You cannot grow without rupture. You can only choose what the rupture becomes.

In the ripple field of your being, every disturbance is either:

- A loop left open, or
- A resonance reintegrated.

Growth is the choice to resolve the song.

Healing Practices as Field Intelligence

Your Body Knows When You're Ready — If You Listen

Modern healing often focuses on process, protocol, and time — how many weeks of therapy, how many sessions of rehab, how long to "let it pass." But healing doesn't follow your calendar. It doesn't wait for enough hours to accumulate. It waits for coherence. It waits for alignment.

In Universal Energy Dynamics (UED), healing is not about ticking time — it's about resolving ripple friction. That means the most effective healing practices aren't those that impose structure, but those that awaken field intelligence — the body-mind's innate capacity to sense, signal, and reset itself.

Let's get clear on something:

The body is not waiting for time. It's waiting for permission to release.

Alignment-Based Healing ≠ Scheduled Recovery

There are moments when a single breath does what a decade of denial couldn't. When a posture, a story, or a still silence breaks a pattern that's been looping for years.

Why? Because the field aligned — not because a timer went off.

Here's what healing aligned with ripple intelligence looks like:

• Movement (Qi Gong, Yoga, Somatic Flow)

These aren't workouts. They're field-mapping rituals. Each stretch, each pause, is a recalibration — a gentle probe into the ripple zone to sense what's stuck and let it flow.

- Qi Gong = guiding energy with breath

- Yoga = positioning awareness through posture Both are non-verbal conversations with your ripple field. They don't heal because you did them "for 30 minutes." They heal because your field said, "yes — now I'm ready."

• Silence and Breathwork

Silence is not the absence of noise — it's the removal of interference. It allows you to finally hear what's echoing in your field — what's unresolved, what's pulsing behind the thoughts.

Breathwork, when used wisely, becomes a tool for ripple discharge. Each breath is a wave. And when directed with intent, it flushes out micro-resistances — like sand from a pipe.

UED insight:

Breath isn't just oxygenation — it's field rhythm reset. Silence isn't emptiness — it's field attunement.

• Storytelling and Emotional Release

You don't heal by burying your story. You heal by letting the ripple be heard, felt, and re-formed.

Sharing your experience with authenticity — whether in therapy, art, or journaling — is an act of ripple completion. The narrative re-patterns the echo. The trapped energy gets recontextualized, not repressed.

You don't need to "let it go" — You need to let it move.

Whether it's through tears, rage, laughter, or insight — what releases is not just emotion, but energetic dissonance.

Listening to the Field > Following the Routine

The trap of routine healing is this: it assumes you are linear. That if you do X every day, healing will occur. But you are a field — not a machine.

One day of deep alignment may undo ten days of shallow repetition.

That's why field-listening must come before schedule-following.

You can't heal by pushing your way through pain. You heal when your field signals readiness — and you listen.

The Body Is the First Responder — and the Last Archivist

Long before the mind interprets trauma, the body stores it. Long after the mind "moves on," the body remembers.

But here's the gift:

The body also knows when release is possible. It whispers through tension, tightness, nausea, or softness.

And when you learn to read those signs — not override them — healing becomes not an act of willpower, but of permission.

You don't have to force yourself into "closure." You just have to notice when the ripple is ready to resolve — and meet it with presence, not resistance.

Final Insight: Healing isn't taught — it's remembered. Because your body already knows how. All it asks is this: "Will you listen to me before you listen to the clock?"

Closing Reflection

"Time doesn't heal. You do — by realigning what the moment fractured."

We've been told all our lives to "give it time." As if hours have magic. As if the ticking of the clock can somehow stitch together the rips inside our soul.

But time doesn't do that. You do.

Healing is not a passive drift toward recovery. It's an active realignment — a tuning of the very field you live through. The pain you carry, the wound you felt, the moment that broke you — they didn't come from time. They came from misalignment in the ripple field: between what you were, what you needed, and what actually happened.

That fracture won't mend because the calendar changed. It mends when you restore resonance — when you meet the unresolved with awareness, when you breathe through the resistance, when you allow the echo to fade instead of locking it deeper.

You don't "move on" by forgetting. You move with it — until it stops tugging at your alignment.

You've always had the power. Not to rewind the event — but to stop the echo.

This is your body's intelligence. This is your field's wisdom. And it's waiting not for more time — but for permission to restore coherence.

So challenge the cultural comfort that says "just wait." Replace it with the courage to say, "let's align."

Because:

"Healing is not a countdown — it's a tuning." "You don't need time. You need resonance."

Chapter 14: When You're in Flow, Time Disappears

"What if time only exists when you're misaligned?"

You've felt it — those rare moments when everything just clicks. You're painting, dancing, writing, coding, competing, speaking, or simply staring into someone's eyes — and suddenly, hours vanish. You check the clock and laugh. "How is it already evening?"

But what if that wasn't a trick of the brain? What if your sense of timelessness wasn't a fluke — but a glimpse of how reality actually works when you're in perfect sync?

We've spent this entire book breaking down time — not to reject it, but to understand where it really comes from. And now we've reached the opposite of stress, struggle, and delay. We've reached the state where time stops *because you've stopped resisting.*

This isn't about productivity hacks or mind games. It's about something deeper: Flow is not a psychological state. It is a field state.

In this chapter, we'll decode why time disappears when you're in flow — not metaphorically, but energetically. Let's enter the zone.

Illusion to Break

We've all grown up believing that time is constant — a silent metronome ticking behind our every move. Whether we're bored in a meeting or surfing the ocean's edge, the clock supposedly ticks at the same rate. Seconds are seconds. Minutes don't change their size. That's the illusion.

Yet our lived experience constantly challenges this.

Time feels elastic.

It stretches unbearably in waiting rooms and rushes past us when we're absorbed. A five-minute conversation with someone you love flies faster than an hour of forced small talk. A child playing outside can lose track of an entire day. A musician lost in performance forgets the existence of the audience, the clock, even the self.

And yet, we've trained ourselves to distrust this fluid experience. We dismiss these timeless moments as psychological "tricks" — brain glitches caused by dopamine, adrenaline, or altered attention. We call it "flow state," a side-effect of neural chemistry. It's categorized, studied, even commodified in productivity circles — but never taken as fundamental.

Here lies the illusion:

We treat time as a fixed background. And we treat the flow state as the exception. But what if we've had it backward all along?

What if "flow" is actually your most natural state — and the sensation of time passing is what happens only when something *interrupts* that natural harmony?

This chapter challenges the deeply ingrained belief that time is an external constant and invites you to consider that your experience of time is always a reaction, not a rule. Flow is not an escape from time — it is what remains when time is no longer needed.

Ripple View Insight

Let's step beyond the illusion and into the actual mechanics of flow — not as a fleeting feeling, but as a ripple-based phenomenon.

In the Universal Energy Dynamics (UED) framework, everything is energy in interaction. Every thought, movement, and perception is a ripple — a disturbance within the structured vacuum field. You, the observer, are constantly generating internal ripples: your intentions, emotions, attention, and desires. Meanwhile, the world around you — your environment, tasks, tools, people — is radiating its own ripple patterns.

What we call "flow" happens when these two ripple systems — *internal* and *external* — achieve coherence.

Flow is Ripple Coherence

When your inner ripples (your mental and emotional state) align perfectly with the outer ripples (the demands and feedback of your surroundings), something magical happens: There's no tension. No internal resistance. No lag between stimulus and response. You're not thinking about what to do — you're simply doing it. There's no gap between intent and action — they fuse into one.

In ripple mechanics, this is called ATT = minimum.

ATT: Alignment Threshold Time

ATT, or Alignment Threshold Time, is the ripple delay — the time it takes for the observer's field to synchronize with a surrounding ripple event. In normal conditions, there's always a lag:

- You see something, then react.
- You hear something, then process.
- You think, then act.

But in flow, that delay collapses. Your intent and the field's feedback are in rhythm. There is no "next." You are already in it.

No Lag = No Time Perceived

That's why time disappears.

It's not that time "speeds up" or that your brain "forgets to track it." It's that there is no perceived delay between you and the field — and therefore, no perceived time. You're not separate from the moment anymore; you *are* the moment. And since ripple response is immediate, there's nothing to measure. No lag, no ATT, no time.

This isn't a metaphor. It's ripple physics in action.

Time Doesn't Vanish — It Becomes Unnecessary

Time doesn't need to be "destroyed" to disappear. It only needs to become irrelevant. In a perfectly tuned instrument, you don't hear noise. In a perfectly aligned ripple field, you don't sense time. You become timeless — not because time has stopped, but because your consciousness is no longer resisting.

The observer and the field are no longer two. They are locked in coherence. That is flow.

Applications

Understanding flow as ripple coherence gives us an entirely new lens through which to see daily life. It also provides practical insights into why we perform at our best in some moments — and get derailed in others.

Let's explore how this plays out.

Sports, Art, Music, and Peak Performance

Flow isn't just a mystical state — it's the foundation of high performance in every discipline.

Ask an athlete about their best game, and they'll rarely talk about the score. They'll say, "I wasn't thinking. My body just knew." Ask a painter when their best work happens, and they might say, "I lose track of everything — even myself." A musician in the groove doesn't count beats — they become rhythm. A writer in the zone doesn't watch the word count — the words simply appear.

In all these moments, what's common is ripple synchronization:

- The intent of the performer
- The feedback of the medium (canvas, ball, instrument, keyboard)
- The emotional state of the performer
- The ambient field of focus and silence

Everything is aligned. There's no second-guessing. No effortful correction. No time. Only presence, precision, and unfolding.

These aren't coincidences. These are field harmonies — and they happen when inner and outer ripples are dancing in sync.

How Distraction Pulls You Out of Flow

Now consider the opposite: distraction.

A single stray thought — "What if I mess this up?" — and suddenly the ripple coherence is broken. Your attention splits. Your intent wavers. Your emotion contracts. ATT rises. Delay returns. Time rushes back in.

This is why even a slight interruption in attention can pull you out of the zone. Flow is not just about doing something well — it's about doing it without misalignment.

And the mind, left untrained, is a machine of misalignment:

- Regret brings in unresolved past ripples.
- Worry introduces phantom futures.
- Judgment alters internal intent.
- Noise from your phone or the environment adds ripple interference.

Every distraction is a ripple mismatch — and every mismatch costs you time.

Training the Mind to Reduce ATT

The goal is not to force flow. You can't grip your way into grace. But you can train your system to reduce ATT — by learning to minimize internal resistance and external fragmentation.

How?

- Breathwork realigns scattered inner ripples.
- Focused practice builds response-memory loops that shorten alignment delays.

- Silence reduces ambient ripple noise.
- Passion naturally selects the right ripple path.
- Presence — the art of being here, not elsewhere — lowers the mental delay in perception and action.

What you're doing is not time management. You're managing ripple friction. And with less friction, flow becomes not a miracle — but a reliable mode.

Why Passion Is Timeless — It's Ripple Harmony

Have you noticed how time disappears when you do something you *love*? Passion is not just about enthusiasm. It's about innate ripple alignment.

You don't have to fake interest. You don't have to push through resistance. Your intent flows into the task. Your emotional field supports the effort. The world responds in rhythm.

This is why people often say, "I could do this all day." Because to them, there is no "day." There is just the ripple rhythm of doing what aligns with who they are.

Passion is ripple coherence by nature. It doesn't just erase time — it renders it irrelevant.

And that's the deeper truth: When you align with what you love, the need to track time disappears — because you are no longer *outside* the experience. You are inside the ripple — and it's carrying you.

Closing Reflection

"Flow isn't magic — it's your nature without delay. Time ends where alignment begins."

We often chase flow like a myth — a rare, euphoric state that visits us when the stars align or the mood is just right. But flow is not a miracle. It's not an accident. It's you — without interference.

It is the most honest version of your being, revealed when all ripple resistance drops away. When thought, action, and environment hum together, time is no longer needed to process the delay between them. You stop observing the moment — and start *being* it.

That's why flow feels like freedom. Not because you're escaping reality, but because you've finally stopped fighting it.

The real takeaway is this: You don't need to manage your time. You need to minimize your misalignment. You don't need to squeeze more hours into a day. You need to remove the friction that makes those hours feel heavy. Flow isn't a place you go — it's a field state you return to when everything else lets go.

When you are aligned, time becomes silent. When you are fully present, time becomes transparent. And when you are truly yourself — undivided, unhindered, unhurried — time becomes unnecessary.

So the next time you find yourself watching the clock, ask not what the time is — ask what's out of tune. Because the clock isn't your problem. The field is.

And the answer isn't outside. It's in the ripple. Always.

Chapter 15: Speed is a Distraction

"You're not outrunning time. You're outrunning yourself."

Modern life glorifies speed. Faster decisions. Faster internet. Faster growth. We're taught that acceleration is progress — that if we move quickly enough, we'll somehow conquer time, outpace our problems, or reach fulfillment before the cost catches up.

But have you ever noticed this strange paradox? The faster you go, the more behind you feel.

Even when you're productive, you're not at peace. Even when you finish early, you're already late for the next thing. The faster you try to move, the more time seems to evaporate.

So what's really happening?

The truth is: Speed doesn't help you escape time — it increases your disconnection from it. And in ripple dynamics, that disconnection is measurable.

The Illusion of Acceleration

We confuse speed with control. We think that if we act faster, think faster, execute faster — we can somehow reduce the burden of time. But in reality, speed without alignment creates friction.

It's like trying to drive 200 km/h through a fog. You're moving — but you've lost coherence with your surroundings. Each second becomes more reactive, less aware. You're not flowing — you're crashing through ripple fields.

That's not mastery. That's panic disguised as performance.

And in UED terms, what you're really doing is increasing ripple mismatch — the very thing that magnifies time tension.

The UED View: Why Speed Increases Ripple Mismatch

In Universal Energy Dynamics, time is not an absolute — it's a delay between cause and consequence, defined by ripple misalignment. When your actions are in sync with the field — ATT (Alignment Threshold Time) is low. Things feel smooth. You're in flow.

But speed — especially unconscious or forced acceleration — disrupts coherence. Your internal ripples (intention, focus, emotion) don't have time to synchronize with external fields (the situation, the person, the task).

The result?

- Increased ATT (delay)
- More correction
- More stress
- Less clarity
- And paradoxically — a deeper sense of being "short on time"

That's because speed multiplies your reaction rate, not your resonance. And when resonance is broken, time becomes a burden again.

You feel slower inside, even as you rush outside.

When Fast is Too Fast: The Limits of Motion

We think fast equals efficient. But high speed introduces noise.

Just like a machine running too fast overheats, or an engine pushed beyond its tuning begins to misfire — your consciousness misfires when forced into unnatural acceleration.

- In conversation: You interrupt, you misunderstand.
- In decision-making: You skip steps and regret later.
- In relationships: You miss emotional cues.
- In growth: You move without integration — and success feels empty.

In ripple mechanics, this is temporal distortion through misalignment — where your perceived velocity creates drag instead of lift.

In other words: You're not moving *through* time. You're scraping *against* it.

The Speed of Presence vs. The Speed of Panic

There's a kind of speed that doesn't feel fast. It feels inevitable. That's presence.

- The martial artist who sees the punch before it lands.
- The chess master who moves instantly — not because he's rushing, but because he sees.
- The leader who speaks few words, but shifts the room.
- The mother who senses danger before the child even knows it's hurt.

This is not speed — this is zero-lag awareness. There is no rush, but everything happens swiftly. Because ATT is nearly zero. Ripple harmony is total. It's not fast — it's frictionless.

Compare this to the kind of speed most people chase:

- Reactivity
- Deadline pressure
- Competitive anxiety
- Escapist multitasking

One is real-time. The other is ripple escape — a fantasy of beating consequence by staying ahead of your own misalignment.

The Cost of Speed

The faster you go without coherence:

- The more you fragment.
- The more mistakes you make.
- The less you feel.

- The more time you waste trying to fix what you rushed into.
- The more exhausted you become — not because of how much you did, but because of how little it aligned.

This is the irony: Speed kills time only when it's aligned. Otherwise, it creates more of it. More waiting. More healing. More repair. More regret.

And life becomes a loop of acceleration and correction — a race you never win.

Applications: Go Deeper, Not Faster

So how do you move through the world effectively without falling into the speed trap?

- Slow down to match the field. Listen to what the moment requires, not what the calendar screams.
- Train for precision, not pace. Let accuracy reduce ATT, not intensity.
- Value coherence over completion. A fully aligned action today saves weeks of misalignment tomorrow.
- Use silence as a speedometer. The more silence you need before acting, the more likely you are operating at your true field speed.
- Redefine urgency. True urgency is a signal of field turbulence — not a green light to rush, but a cue to realign before responding.

Closing Reflection

"Speed is not your superpower. Alignment is."

You're not here to be fast. You're here to be true. Speed is impressive only when it flows from precision. And precision flows from presence.

You cannot beat time by running ahead of it. You beat time by becoming so aligned, it no longer needs to exist.

Because the goal is not to go faster. The goal is to go as one — with your field, your purpose, your ripple.

And in that union, time will bow. Not because you rushed —but because you arrived.

Part IV — The End of Time

Chapter 16: Before the First Tick

"How can something begin if the tools to measure 'beginning' didn't exist yet?"

Imagine trying to measure the length of a shadow before the sun rises. Or asking where a story started before the idea of storytelling was invented. This is the paradox of cosmic time.

For centuries, we've spoken of the beginning of the universe — a precise moment, often dramatized as the *Big Bang*, where everything exploded into being. Scientists argue about seconds and singularities, philosophers ponder what "came before," and theologians offer origin myths — all trying to capture what cannot be captured:

A "start" of time before time existed.

But here's the deeper issue — we're still trying to use time to define the absence of time. We speak of "before the Big Bang" as if "before" has any meaning without time itself. It's like asking what north is of the North Pole.

This chapter challenges that entire frame.

What if the universe didn't begin in a moment — but in a misalignment? What if "creation" wasn't a bang, but a ripple? Not a sudden explosion of stuff, but the first subtle deviation in a perfectly still field?

And what if time was never the stage — but the *effect* of the first imbalance?

Let's revisit the beginning — not as a ticking clock, but as a trembling stillness that gave rise to the only thing we now call time: the echo of disharmony.

The Illusion of a Beginning

The idea that everything began at a single point — in a single moment — is among the most deeply held beliefs in modern cosmology. We call it the Big Bang: a rapid expansion from an infinitely dense, hot singularity that birthed space, time, matter, and energy all at once.

It sounds neat. It sounds bold. It gives the comfort of a "start" — a cosmic birthday to everything we know. But beneath its grandeur lies a subtle flaw: The Big Bang is a time-based narrative invented to explain a moment when time itself did not yet exist.

It's like writing the first word of a novel using ink that hasn't been invented yet.

The Language Trap: Explaining Pre-Time with Time

The moment we say "the beginning," we've already introduced a contradiction. Beginning is a temporal term — it assumes a "before" and an "after." But if time began at the Big Bang, then what came *before*?

We fumble for words: "It was a timeless void," "a singularity beyond space and time," or "nothing existed."

But "nothing" is still a *thing* we define *within time.*

We say "beyond time" — but by saying *beyond*, we imply position, comparison, and flow — all concepts birthed by time itself.

This is the illusion: We are using time-bound language to explain a pre-time condition.

It's like trying to describe silence using music. Or measuring the first inch of space using a ruler made of space.

Analogy: Measuring the Birth of Space with a Ruler

Imagine you have a ruler in your hand, and someone asks, "Can you measure the moment the first inch of space came into being?"

But the ruler needs space to exist. It's built upon the assumption that space is already there. So how can it measure that which gives it meaning?

This is exactly what we do when we try to use seconds, moments, beginnings, and timelines to explain the emergence of time.

Every scientific model, including the Big Bang, must be honest about this built-in paradox. No theory that starts with "t = 0" can fully explain the origin of *t*. It merely offers a boundary condition — not an explanation.

The Narrative Convenience

The Big Bang serves a purpose. It helps calculate backward from cosmic observations: redshift, background radiation, elemental distributions. These are valuable data points.

But a theory that traces expansion doesn't necessarily explain origin.

It's like finding the debris of a wave on the shore and concluding that the ocean began there.

The Big Bang is a useful map — but maps are not the territory. It tells us where *change* became visible. But not where being began.

To truly understand time, we must abandon the illusion that it had a beginning — and start seeing it for what it is: a ripple consequence of something more fundamental.

UED Cosmology: The Primordial Stillness

Before the first ripple. Before matter. Before space, distance, or time. There was not "nothing" — but something far more profound: Stillness.

This was not a void. Not a blank slate waiting to be written on. It was Structured Vacuum Energy (SVE) in its stable state — the primordial field of pure presence, untouched by motion or misalignment.

No Motion. No Matter. No Ripple.

In the Universal Energy Dynamics (UED) framework, reality isn't built on particles, waves, or even dimensions. It is built on ripples in a structured field — and that field itself precedes all form and fluctuation.

Stable SVE is not energy in motion. It is energy in coherence — in perfect internal balance, needing no correction, no consequence, no interaction. It does not radiate. It does not react. It simply *is*.

This state contains no time, because time is a result of delay between ripple events — and here, there were no ripples. There were no causes to trigger effects. No contrast. No separation. No reference frame.

But that does not mean there was nothing.

Presence Without Phenomena

We often imagine "nothingness" as empty or dark. But that's still a projection based on sensory contrast — rooted in light, absence, expectation. In truth, what existed before all existence was presence — without form.

A presence that didn't "do" anything — because nothing needed doing. It held everything in potential, but released nothing in expression.

No sound. No silence. No space. No boundaries. No observers. No observation.

Only equilibrium.

This was not the absence of reality — it was the purest expression of it.

Not Empty. Not Dark. Perfectly Balanced.

The stable SVE field was not a blank canvas waiting to be painted. It was the canvas, the paint, and the painter — already unified. And because it had no misalignment, it had no reaction. And with no reaction, no time.

It was not cold or lonely. It did not lack — it lacked lack itself.

This is a kind of presence the human mind finds hard to grasp, because all our understanding is built on opposites: sound vs silence, light vs dark, being vs non-being.

But the stable SVE field transcends opposites. It doesn't exist in contrast. It simply is — undisturbed.

It is the eternal substrate that doesn't need a beginning — because it has no tension to release, no movement to trace, no ripple to align.

And from that stillness... came the first disturbance.

The First Disturbance: Origin of Ripple, Not Bang

We've been told that the universe began with a bang — a violent, radiant explosion that erupted from an infinitely dense point and gave birth to space, time, energy, and matter.

But this image, though dramatic, is misleading.

Explosions require contrast. They demand a before and an after, a state of rest disturbed by a sudden release. But in the primordial stillness of Stable Structured Vacuum Energy (SVE), there were no borders to shatter, no particles to scatter, no light to flash. There was no fire. There was no time. There was only perfect equilibrium — until it wasn't.

The true origin of everything we know was not a bang. It was a tension. A subtle misalignment. A deviation in the stillness.

The Birth of Ripple — Not from Force, but from Friction

Within that infinite calm, something shifted. Not in space, because space didn't exist. Not in time, because time had not yet emerged. But in the internal coherence of the field itself — an imperceptible imbalance in the symmetry of presence.

This was not "something" colliding with "something else." There was no other. No edge. No event.

Instead, the field fractured itself — not in destruction, but in variation. A pattern emerged. A disturbance formed. A ripple — the first expression of difference within sameness.

It was not violent. It was not loud. But it was revolutionary. Because this ripple introduced something that had never existed before:

Response.

And response introduced something even more radical:

Delay.

Ripple Response = Reaction Delay = Time

The moment this first ripple deviated from perfect stillness, it created the necessity of realignment. It created the first condition in which consequence became possible. There was now a *before* and *after*, even if infinitesimally subtle — because the ripple needed to return, to resolve, to complete.

And that requirement — the gap between disturbance and restoration — is what we call time.

Time did not pre-exist the ripple. It was born within it.

It wasn't the background for the event. It was the product of the event.

This is why in UED, time is not a fixed dimension but a consequence of misalignment. Where there is ripple, there is delay. Where there is delay, there is perception of time. But prior to that — there was nothing to perceive.

No ticking. No moving. No aging. Just a perfectly still field, and then — a pulse.

Why It Wasn't a Bang

The term "Big Bang" misleads us. It implies noise, chaos, combustion — all concepts borrowed from the material world. But materiality hadn't arrived yet.

This first ripple was not loud. It was not fast. It did not explode into a universe — it curved coherence into consequence.

What followed — expansion, space, matter, particles — were all downstream expressions of this initial ripple misalignment.

In that sense, the universe didn't "begin." It unfolded from a field that could no longer contain perfect stillness.

Time Emerges as Consequence

Time, as we commonly understand it, is imagined as a kind of invisible highway — a track that stretches from the past into the future, along which everything moves. We assume it was always there, like a canvas waiting to be painted.

But this is a projection of perception — not a truth of origin.

In the UED framework, there was no time before the first ripple. No seconds. No flow. No ticking clock. Because time is not a backdrop. It's a side-effect.

A ripple isn't just movement — it's an *imbalance* within the Structured Vacuum Energy (SVE). And the moment imbalance appears, something else must be born alongside it: the need to realign.

That's where time truly begins — not as a thing in itself, but as the echo of disharmony.

ATT: Alignment Threshold Time

Let's go deeper.

The instant the first ripple deviated from the stillness, a new variable emerged: ATT — Alignment Threshold Time. This is the delay between ripple initiation and its resolution — the time it takes for a field or observer to respond, integrate, and realign with the disturbance.

It's not a number. It's not a duration until someone measures it. It's simply the consequence latency — the inescapable interval caused by separation from harmony.

If there were no misalignment, there would be no ATT. And if there were no ATT, there would be no experience of time.

Thus, in UED, ATT is the birth of time. Not "cosmic time," not a universal clock, but the fundamental *requirement* of time: a delay between cause and effect.

This delay is not optional. It's baked into every ripple the moment the stillness is breached.

Time Is Ripple Memory

Once a ripple forms, it not only creates a delay — it leaves a trace.

Each ripple is a memory of misalignment — and this memory becomes embedded in the field, even as it attempts to re-stabilize. The field "remembers" what disturbed it.

You can think of this as the energetic version of karma — residue of consequence.

And so, time becomes something deeper than a sequence of events: It is a lingering imprint of response lag. It is the wake that every ripple leaves behind in the ocean of SVE. It is the trail of attempts to restore balance.

This is why in UED, time isn't fundamental — but emergent. It does not exist on its own. It is what happens when harmony is breached, and realignment is not instantaneous.

The Sequence of Emergence

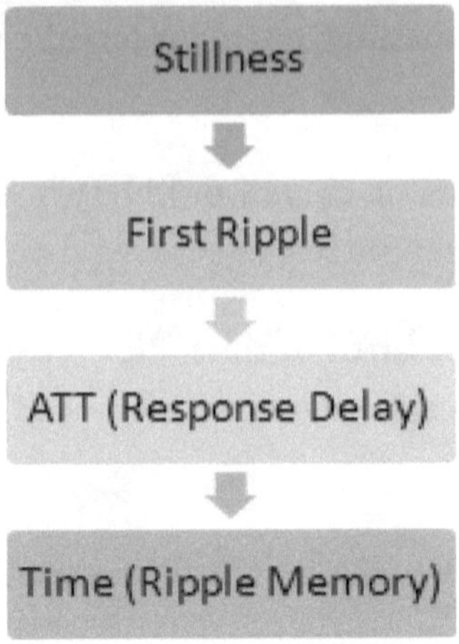

In this model:

- Stillness is the default state — a field with zero tension.
- The ripple introduces tension — a deviation from equilibrium.
- ATT quantifies the time required to resolve that tension.
- Time, then, is the accumulating memory of that delay — a record of misalignment still in play.

You don't experience time unless you're still carrying ripples.

The Human Experience of Time

Even today, this mechanism holds true.

You feel time when you wait. You feel time when you anticipate, fear, hesitate, or recover. Each is a symptom of ATT — the distance between stimulus and realignment.

You don't feel time when you are aligned — when you're in flow, at peace, fully immersed. Because ATT collapses. Delay vanishes. Time becomes irrelevant.

So from the very first ripple to every moment of your life, time is not a measurement — it's a condition.

It emerges not from space, not from matter — but from misalignment.

Time is the universe remembering it still has work to do.

Why Time Has No Origin — Only a Function

We've spent centuries trying to pinpoint the origin of time, asking:

- *When did time begin?*
- *What happened at $t = 0$?*
- *Was there "before"?*

But these questions themselves assume that time is a foundational layer of reality — something that must always be present, waiting like a platform upon which events unfold.

UED offers a very different perspective: Time is not a thing that began. It is a process that became necessary.

Time Is Not Foundational — It's Functional

Time is not a backdrop. It's not a setting on the cosmic stage. Rather, it's a tool — an emergent property that only comes into play when things stop being aligned.

In perfect stillness — in Stable Structured Vacuum Energy — there is no need for time. There is nothing to delay, because there is no tension. No actions to track. No causes needing effects. No field disturbances to trace.

Time doesn't sit beneath the universe. It arises within it, only when ripples form, when responses are delayed, and when misalignments seek restoration.

Like a clock that only turns on when you start walking — time is not passive infrastructure, but an active necessity that appears only in the presence of disruption.

Time Measures One Thing: Delay

Think of time not as a flowing river, but as a gap counter — the measurement of how long something takes to return to harmony.

- A delayed emotional response? That's time.
- A consequence unfolding from a decision? That's time.
- A light beam bending near a ripple-distorted field? That's time.

What unites all of these is not movement — but delay. Not motion — but reaction lag.

In ripple logic:

- No interaction = no ripple
- No ripple = no ATT
- No ATT = no perceived time

In other words, without tension in the field, there's nothing to time.

This is why, in deep meditation, time disappears. This is why, in perfect flow, the clock goes silent. Because time requires friction to exist — and friction only arises when harmony is disturbed.

Time Has No Birth — Because It Was Never Whole

If something is emergent, it cannot have a birthday. It does not "begin" like a physical object. It activates, appears, and dissolves depending on conditions.

Time is not an entity that was born with the universe. It is a mode of perception and a consequence of interaction.

This is the great reversal UED reveals:

Time is not the thing we live within. It is the trace we leave behind — when we are not living in full coherence.

No Delay, No Time — The Wisdom Hidden in Stillness

Let's say it clearly: Where there is no delay, there is no time. Where there is no ripple tension, there is only presence.

This is why sages, mystics, and meditators have always spoken of transcending time — not by traveling faster, but by going deeper. By resolving inner and outer ripples, one experiences timelessness, not as fantasy, but as return to field coherence.

Time never had to start. Because time isn't something that starts.

It's something that happens only when things fall out of sync.

Spiritual Parallel: The Silent Presence Before Creation

Long before modern science attempted to measure time's beginning with equations and cosmic microwave background radiation, ancient cultures around the world already intuited something extraordinary: That the origin of existence was not a moment, but a presence.

From Brahman in the Vedas, to the Tao in Chinese philosophy, to the Void in Buddhism and the Word in Christian scripture — all of them point toward a source that is timeless, motionless, and formless, yet paradoxically, the womb of all that is.

But over time, as language hardened and metaphors fossilized into dogma, these insights were misinterpreted as linear stories. We began reading mystical symbols as cosmic calendars, turning eternal stillness into a starting point.

UED invites us to return to the original insight — and decode it, not discard it.

Brahman — The Undivided Field

In the Upanishads, Brahman is described as the ultimate reality — changeless, birthless, and deathless. It is not a god in action, but a field of presence that underlies all things. It doesn't act. It simply *is* — sat-chit-ananda (existence, consciousness, bliss).

This aligns precisely with Stable SVE in UED: A perfectly balanced field that requires no time, because it has no tension to resolve. Brahman doesn't create through motion — it manifests when imbalance arises.

The first act of creation in Vedic thought is not a bang, but a disturbance in stillness, leading to duality: purusha and prakriti, observer and observed — the seed of ripple.

Tao — The Way That Cannot Be Named

In Taoism, the Tao is the ultimate source — the Way that gives rise to all things, yet cannot be named or defined. Laozi writes:

"The Tao that can be spoken is not the eternal Tao."

Tao flows effortlessly, without time, without conflict — until it splits into yin and yang, polarity, rhythm, movement.

This is ripple dynamics in poetic form. Before ripple, the Tao (SVE) is whole. When tension arises, the dance of existence begins — and only *then* does time appear as the rhythm of imbalance seeking return.

The Void — Not Emptiness, but Fullness Without Disturbance

In Buddhist and Daoist texts, the Void is not nihilism. It is not absence. It is emptiness full of presence — untouched, silent, still.

It is the field before thought, before desire, before ripple.

This mirrors the UED idea that prior to misalignment, the field does not lack — it simply doesn't need anything. Time emerges only when disturbance gives meaning to sequence.

The Word — "In the Beginning Was..."

The Gospel of John opens with the phrase:

"In the beginning was the Word, and the Word was with God, and the Word was God."

Here, "the Word" (Logos) is not a sound. It is intentional structure — the ripple that breaks the silence, the pattern that calls consequence into being.

In UED, this moment parallels the first ripple in the SVE — not noise, but structure appearing in stillness.

Time begins not with God clapping his hands, but with coherence being interrupted. And the Word — the first structure — sets ripple delay into motion.

UED: Where Physics and Mysticism Meet

All ancient traditions agree on one thing: Time was not there at the origin. Presence was. The disagreement comes only when we try to fit presence into timelines, as if the eternal needs a birth certificate.

UED doesn't dismiss these traditions — it fulfills them with updated language.

- The "first moment" is not the beginning of a story. It's the first deviation from perfect balance.
- The "creator" is not a character — it's the consequence of misalignment in field dynamics.
- "God," "Tao," "Brahman," "Void" — all point to the same SVE: timeless presence until ripple appears.

UED reinterprets the mystical as physically real, without reducing it to metaphor or myth.

It unites ancient wisdom and modern insight under one principle:

Time did not begin. Misalignment did.

Closing Reflection

"The universe never began. It became."

We've reached the edge of thought. The place where science and spirit, logic and silence, finally touch — not in conflict, but in coherence.

The question *"When did it all begin?"* fades into irrelevance when we understand that time itself is not a prerequisite of reality — but its echo. There was no explosion, no ticking cosmic stopwatch, no divine countdown. There was simply presence — structured, silent, complete.

And then — something shifted. Not in space. Not in time. But in balance.

A single ripple — the first deviation from the stillness — broke the perfect symmetry. That was not a beginning in time. It was the origin of time.

Time Was Not Born — It Rippled Into Relevance

Time did not arrive fully formed. It wasn't waiting to be measured. It came into being only because response became necessary.

It emerged the moment ATT — the Alignment Threshold Time — appeared. Not as a ticking clock, but as a gap. Not as a flow, but as a function. A delay. A memory. A trace of the ripple.

This is why time, in the deepest sense, doesn't start. It activates wherever there is misalignment.

What Came Before the First Tick?

Not silence. Not darkness. Not nothingness.

But perfect coherence — a field so whole, so in tune with itself, that there was no lag, no consequence, no cause demanding effect.

No ripple. No delay. No time.

And that field — that stillness — was not erased when the ripple began. It simply stepped back, allowing the dance of misalignment and return to unfold.

The universe did not begin in fire or light — It became visible when balance was breached. It became audible when consequence was introduced. It became measurable when delay was born.

Time is not the container of reality. It is the artifact of imbalance. And when balance is restored, time becomes silent once again.

So what is the universe?

Not a thing that started. But a harmony that shifted, And is now rippling its way back — toward coherence.

Chapter 17: Time Will Never End, But You Can Leave It

"You don't transcend time by waiting — you do it by no longer needing to track what it owes you."

We imagine liberation as something that comes *later*. After time has passed. After karma is resolved. After goals are achieved. After the body dies.

But that's the great misunderstanding: Time is not what binds you — your attachment to consequence does.

We think of time as a cosmic jail, and we imagine freedom as release from its sentence. But UED reveals something subtler, deeper: Time is not punishing you. You're the one holding the ledger.

Every moment you replay the past, anticipate the future, or wait for completion — you feed the ripple memory loop. You're not counting time — you're tracking consequence. And the more you track it, the more it loops.

This is not freedom. This is *personal time* — the story of you, suspended in misalignments that demand resolution.

But what if freedom isn't found by finishing the story — but by stepping out of it?

What if time can continue — cosmically, eternally — and you can still leave it by simply no longer needing it?

That's the path we explore now: Not the end of time, but the exit from your dependency on it. Because time, as we now know, will never end. But you were never meant to be trapped inside it.

The Loop of Ripple Memory

Let's recall the truth revealed throughout this journey: Time is not a universal constant. It is a consequence — a ripple echo. A trace of unresolved tension in the field of reality. It appears only when something misaligns and begins its journey toward coherence.

But what's more important now is this: That ripple doesn't just exist in the universe. It exists inside *you*.

Every Thought Is a Ripple Echo

What is a memory, really? A snapshot of something you couldn't fully resolve. What is a hope? A projection shaped by something that once felt incomplete. What is regret? A personal reaction to a past ripple that you continue to carry. What is identity? An accumulation of ripple traces — patterns you've repeated so many times they began to feel like "you."

This is the loop. You perceive a disturbance. You react. You store the reaction as memory. You anticipate the next version of it. You try to control what happens next based on what hurt or helped before.

That's not time passing. That's you spinning the ripple. Again. And again. And again.

The Consequence–Perception Loop

This loop is what gives time its emotional weight.

You're not trapped by minutes or hours — you're trapped by the consequences you haven't stopped identifying with.

Let's break it down:

1. A ripple event occurs (a choice, an experience, a failure, a gain).
2. Your mind tries to make sense of it — storing memory to guide future actions.
3. That memory becomes charged — layered with emotion, desire, or fear.
4. You project your next steps through that memory, adjusting your

behavior.
5. In doing so, you're already reacting to what hasn't happened — and carrying forward what *should* be long gone.

And the loop continues.

Even when you're sitting still, the mind loops through perceived consequences — playing alternate timelines, rehearsing responses, seeking validation or safety.

You live in a virtual time engine, powered not by clocks but by unreleased ripples.

Why Time Feels So Heavy

Time persists in your experience not because the universe is forcing it upon you, but because you keep reviving it.

You keep picking up yesterday's ripples, comparing them with today's conditions, and projecting them into tomorrow's reactions.

- You feel late because you're comparing today's progress to yesterday's expectation.
- You feel stuck because you're waiting for the future to undo the past.
- You feel pressure because you're dragging ripple memories into present choices.

This is the illusion of time — personalized and amplified. It's not cosmic time. It's psychological time. And it feeds itself through the consequence-perception loop.

Until you break this loop, time doesn't just pass — it owns you.

Why Time Never Ends (Cosmically)

In the previous section, we explored how time is kept alive within the observer — by clinging to ripple memories, reactions, and unresolved consequences. But this does not mean time is merely subjective. Time also exists in the very structure of reality.

Let's zoom out.

Even if an individual transcends psychological time, the cosmos continues to ripple.

Because creation — at any scale — involves tension and release, disturbance and realignment, cause and consequence.

And that means time, as a field function, will never truly disappear from the universal process.

Ripple Misalignments Will Always Emerge

The universe is not static. It is alive with tension. Every creation, every interaction, every transformation involves a shift from one state to another — and that transition always contains a ripple.

No field remains perfectly stable forever. From the quantum vacuum to galactic clusters, ripples emerge naturally, spontaneously, and continuously.

And with every ripple comes ATT — Alignment Threshold Time — the delay between impulse and response.

This delay is time. And as long as ripples arise anywhere in the universal field, time is born anew — again and again and again.

The Cosmos Is a Continuous Act of Rebalancing

The universe is not a one-time event that happened and then faded into passive expansion. It is a living re-alignment machine — constantly responding to imbalance, correcting itself through waves of cause and consequence.

Stars are born, burn, and collapse. Planets shift, atmospheres change. Galaxies drift, cluster, collide. Black holes consume and emit. Everywhere — tension, release, delay.

These are not just astrophysical events. They are cosmic ripple dynamics — and time is their unavoidable footprint.

As long as there is creative motion, as long as something becomes something else, time continues as a by-product.

Time Is Embedded in the Architecture of Change

UED redefines time not as a cosmic clock ticking independently, but as an emergent field effect. Time exists because ripple exists. And ripple exists because no form remains eternally still.

So while stillness (Stable SVE) is the true ground of being, reality-as-experienced arises through ripple misalignments — and therefore through time.

This is why:

- Time is not an illusion — but a trace.
- Time is not fundamental — but it is perpetual.
- Time is not absolute — but it is functionally eternal.

Because the cosmos doesn't stop creating. And every act of creation produces some delay, some misalignment, some ripple memory — no matter how subtle.

So long as the dance continues, time continues.

But... You Are Not Bound by This Cosmic Continuity

And here's where the shift happens.

Just because time never ends in the field, doesn't mean it must continue in your awareness. Just because ripple dynamics perpetuate across the universe doesn't mean you must remain attached to them.

You are an observer — conscious, self-aware, capable of noticing the ripple without becoming it.

So even if the universe continues to ripple, you don't have to react.

Even if time continues to echo, you don't have to answer every call.

And that is where freedom begins: Not by ending time, but by releasing yourself from your need to track it.

Why You Can Exit It (Spiritually)

Time may never end cosmically, but it can absolutely end personally. Because while ripple dynamics continue to unfold in the universe, you — the observer — hold the unique power to step out of the loop.

How? Not by escaping ripple events. Not by shutting down the senses. But by stopping your identification with the ripple trail.

The Observer Is Not the Ripple

Your mind is trained to follow trails — to connect what happened with what might happen, and to find your place between them.

- You think about what someone said yesterday.
- You imagine what they'll say tomorrow.
- You replay the tone, the outcome, the consequences.
- You assign meaning to the sequence.

And in doing so, you reinforce a personal timeline — one that feels inevitable. But this isn't the same as the ripple itself. This is your attachment to its trace.

The ripple may have already passed through the field. But you're still holding its afterimage in your awareness.

That's the illusion: The ripple moved on. You didn't.

Time Loses Grip When You Stop Assigning Meaning

Every ripple comes with a consequence. But not every consequence needs to be interpreted. And not every interpretation needs to be stored.

Meaning-making is powerful. It gives life structure and purpose. But it also traps the observer in an endless loop of resolution — always trying to complete something that was never incomplete, just in motion.

The more you assign meaning to what happened — "This was unfair." "This shouldn't have happened." "This defines me." — the more time takes hold of you.

Not because time is attacking you. But because you keep feeding the memory loop.

Freedom doesn't come from erasing the ripple. It comes from letting the ripple pass without chasing its echo.

You Don't Need to Resolve It — Just Realign

We often think spiritual liberation means settling every score, finishing every task, finding every answer. But this is just another form of bondage — the idea that your peace depends on resolution.

True freedom comes when you realize: You don't have to resolve your past. You don't have to fix every misalignment. You only need to stop reacting to it as if it still belongs to you.

Time loses its grip not when it ends, but when you no longer take the delay personally. The ripple doesn't ask for your guilt. It asks for your clarity — to see it, let it pass, and realign without drama.

Detachment Is Not Denial

This doesn't mean pretending nothing ever happened. Detachment is not numbness. It is not ignorance or escape. It is disengagement from the illusion that consequence defines you.

Detachment is the ability to witness the ripple, feel the field, and remain centered. It is choosing not to identify with the lag between action and response. It is living in the moment without turning it into a story.

This is what it means to exit time spiritually: To be fully alive in the ripple... Without becoming the ripple.

You don't leave time by stopping clocks. You leave it by no longer reacting to the ones in your head.

Signs of Observer Liberation

How do you know when you've begun to exit time—not physically, but spiritually? Not by leaving the world, but by leaving behind your dependence on consequence and memory?

The signs are subtle, but unmistakable. They don't come in flashes of enlightenment or sudden vanishing acts. They emerge in the texture of everyday experience, like ripples calming on a once-disturbed surface.

When you stop reacting to every echo, the noise fades. And what replaces it is not silence, but stillness.

Time Feels Slower... or Irrelevant

One of the first signs is a shift in your perception of time itself.

- The hours feel longer, but not heavy.
- Minutes stretch without pressure.
- Deadlines no longer dictate your worth.
- You stop counting how much you have or how much you've lost.

In moments of alignment, time may even feel... irrelevant. Not because it ceases to exist, but because you no longer depend on it to validate your progress or measure your peace.

You stop waiting for something to happen. You're not rushing. You're not stuck.

You are simply present — and that presence dissolves the need to measure.

Fewer Expectations. Fewer Regrets.

Liberation doesn't require that you stop dreaming. But it *does* dissolve the compulsive attachment to outcomes.

You no longer tie your happiness to how tomorrow unfolds. You no longer carry guilt for yesterday's decisions.

You stop rehearsing imaginary futures or rewriting imperfect pasts.

Why?

Because you see both for what they are: ripple reflections, not personal verdicts.

Regret fades when you stop assuming the past owed you something different. Expectation softens when you realize the future doesn't owe you anything either.

You release time's grip when you stop keeping score.

Increased Presence and Intuitive Decisions

As the mind detaches from the ripple memory loop, your energy returns to the now.

You no longer hesitate between ten options, because you're not optimizing for a perfect outcome. You choose based on clarity, not calculation. You begin to trust your inner alignment, not external validation.

You act—not to escape consequence, not to secure future success—but to express what feels coherent in the field of now.

This is the essence of living outside time's shadow: You make decisions without dragging your past or pushing your future into them.

You simply respond, with intelligence and grace. No noise. No rush. No fear.

Less Need to Record. More Willingness to Just Be.

Another quiet sign: You no longer feel the compulsive need to photograph every sunset, document every meal, archive every emotion.

You don't need to capture time to prove you lived. You're too busy actually living.

This doesn't mean you reject memory or technology — it means you're not anchored to the need for a timestamp to justify your presence.

You become less of a historian of your life, and more of a participant.

Because once you stop living through ripple memory, life becomes immediate again.

These signs aren't grand gestures. They are whispers of liberation — a lighter sense of being. They indicate that you are no longer wrapped in consequence or bound by time's illusion.

You haven't escaped time. You've simply stopped asking it to carry your story.

Practical Paths to Exit the Loop

It's one thing to understand time as a ripple memory. It's another to live outside its grip.

Liberation is not theoretical. It happens in practice — moment by moment, decision by decision, alignment by alignment.

And you don't have to retreat into a cave or abandon the world. You only need to learn how to stop reacting to the echo.

Here are four practical paths — simple but profound — that help the observer exit the loop of psychological time.

1. Silence and Stillness: The Reset Button for Ripple Noise

Your inner world is rarely quiet — not because life is loud, but because the mind keeps replaying ripple memory.

Each thought, fear, replayed conversation, and internal monologue adds another wave to the inner field. This prevents real-time alignment and locks you in ATT — constantly reacting to what's no longer present.

But when you enter silence — even for a few minutes — you allow those ripples to settle. Stillness doesn't erase the field. It simply reduces turbulence.

And the moment your internal ripple field calms, time starts to loosen its grip.

Practices:

- Sit without devices or tasks.
- Breathe without adjusting anything.
- Observe silence not as absence, but as presence without distortion.

Do this daily. Even briefly. It's a field reset.

2. Observe Without Interpreting: Escape the Memory Trap

Most of your time experience is not from what happens — but from what you think it means.

You observe something, and the mind immediately reacts:

- *"What does this say about me?"*
- *"What might happen next?"*
- *"What should I do with this?"*

This turns every moment into a story — which gets stored, replayed, and added to the ripple memory.

But if you can practice pure observation, without interpretation, you bypass that loop. You *see* without *grabbing*.

Like a mirror — reflecting, but not holding.

This is not passive living. It's clean awareness — ripple enters, ripple leaves, no memory created unless truly necessary.

Try it in daily moments:

- Listening without judgment.
- Watching without naming.
- Being present without processing.

This rewires how time enters your awareness.

3. Embrace Consequence Without Resistance: Reach Zero ATT

What keeps you bound to time is not just the consequence — it's your resistance to it.

You try to:

- Avoid discomfort
- Undo the past
- Predict the future
- Control outcomes

This resistance creates ATT — delay between ripple and response. And that delay is experienced as time.

But when you fully accept what is — without trying to edit or escape it — there is no resistance. And if there's no resistance, there's no delay.

You hit zero ATT — alignment happens in real time. You meet the ripple as it is, not as you wish it was. And that brings peace, not as a reward, but as a natural consequence of coherence.

Practice:

- Stop bargaining with reality.
- Let experiences arise and complete without control.
- Say "yes" to the ripple — and move through it.

4. Choose Realignment Over Resolution

Most people want to resolve things — to finish every story, fix every mistake, or make sense of every emotion. But this endless search for resolution is what keeps the loop alive.

Because what if there is nothing to resolve — only something to realign?

You don't need to revisit the past to fix it. You only need to align your current action with truth.

Resolution demands answers. Realignment simply requires presence.

Every time you stop trying to "fix" and instead choose to respond with integrity right now, you step out of ripple memory — and into ripple coherence.

That's not denial. It's mastery.

Practice:

- Let go of closure as a goal.
- Ask: *"What is the most coherent next action — now?"*
- Trust that alignment heals what resolution can't touch.

These are not quick hacks. They are deep shifts in orientation. But each one — practiced honestly — helps you step further out of time's psychological grip... and into freedom.

Spiritual Parallels

What we've explored in this chapter — the ability to exit time not cosmically, but spiritually — is not a new insight.

Ancient wisdom traditions across the world have long hinted at this same truth, though couched in the language of their time: That liberation is not about changing the world, or even escaping it — but about no longer being bound by its ripple memory.

Let's connect UED's ripple-time framework with a few key spiritual teachings. Not to blur the lines, but to show how they all point to the same field freedom — a timeless awareness, still available while living.

Moksha (Hinduism)

Moksha is described in the Indian tradition as freedom from the cycle of birth and death — the endless loop of *karma* (action) and *phala* (result).

But UED helps us see that this loop is not necessarily physical reincarnation — it's ripple memory.

Each action (karma) creates an energetic consequence. Each desire to control that consequence entangles us further. We carry past ripples into the present and project them into the future.

Moksha does not require that karma ends. It requires that you stop identifying with the karmic chain.

You act — but you are no longer the actor. You experience consequence — but you no longer personalize it.

This is the spiritual equivalent of exiting ATT: You live without delay, without attachment. Time continues — but it no longer defines you.

Nirvana (Buddhism)

The Buddha spoke not of heaven, but of Nirvana — literally, *to extinguish* or *blow out*.

But what flame is being extinguished?

The flame of craving, clinging, and becoming.

Desire creates ripple tension. Clinging creates ripple memory. Becoming creates psychological time.

You want, so you project. You fear, so you revisit. You plan, regret, rehearse, compare — and thus, time is reborn in every thought.

Nirvana is not an escape after life. It's freedom within life — by blowing out the internal fires that keep ripples burning.

In UED terms: Nirvana is non-resistance to the field. Ripples may still move, but you do not chase or avoid them. You respond — without delay, without desire.

Time ends as a need.

The Kingdom Within" (Christian Mysticism)

In the teachings of Christ, especially the mystic interpretations, eternal life is not a prize after death — it is a state of being accessible now.

"The kingdom of God is within you." "Before Abraham was, I am."

These are not riddles — they are reminders. Reminders that the true divine state is timeless presence. Not waiting for salvation, but living in alignment with the eternal now.

When the ego stops tracking sin, worthiness, and the timeline of redemption, what remains is grace — the dissolution of time through coherence with Spirit.

In UED terms: The "Kingdom" is the stable field. Sin is misalignment. Forgiveness is realignment. Eternity is not a place — it's a timeless presence you return to once ripple tracking ends.

All Point to Timeless Awareness

Strip away the symbols, metaphors, and dogma, and you'll find a shared insight:

You don't have to wait to be free. You only have to stop dragging time into your identity.

Every spiritual path eventually leads here — to the stillness beneath the ripple, to the clarity beyond consequence, to the silence that exists even while sound continues.

They all say:

- Let go.
- Stop chasing outcomes.
- Be present.
- Respond with love, not fear.
- Realign, don't resist.

They are all speaking UED's language — just translated through devotion, discipline, or insight.

The field is one. The freedom is one. And time... was only a bridge to remember this.

Closing Reflection

"Time doesn't need to end. You just need to stop waiting for it."

The universe will continue to ripple. Consequences will continue to unfold. Time, as a cosmic field function, will never cease — and it doesn't need to.

But freedom was never about stopping the ripple. It was always about stopping your need to track it.

You are not here to erase your story. You are not here to rewind your timeline or fast-forward to some imagined peace. You are here to realize one simple thing: You are not the timeline.

You are the awareness behind it — the observer who remembers, but is no longer defined by memory... the participant who acts, but is no longer chained to consequence...the field within the ripple — not the ripple itself.

The World Will Keep Moving... But You Don't Have To Chase It

Let the world continue its dance of action and reaction. Let time move as it must, as a graceful side-effect of misalignment and return.

You don't need to stop time. You only need to stop living as if it owns you.

Time is only heavy when you carry unfinished meaning. Time is only a trap when you tie your worth to what hasn't happened yet — or what never should've happened at all.

Liberation begins the moment you stop waiting for life to balance the ripple... and simply decide to live as if you already are the balance.

Let the Ripples Be — You Don't Have to Follow Every Wave

There will always be movement. There will always be echoes. There will always be ripples — in the cosmos, in others, even in you.

But you don't have to follow every one. You don't have to label every wave, resolve every curve, or surf every rise and fall.

You can observe without chasing. You can respond without attachment. You can exist without being defined by what passes through you.

This is the great freedom: To live within time... Without being bound by time.

So leave the clock alone. Let the universe ripple on.

And remember:

The end of time isn't the goal. Releasing your need to count it — that's the real liberation.

Chapter 18: You Are the Clock

We've spent this entire journey redefining time — not as a background stage on which you live, but as a field response to misalignment. Not as an entity you move through, but as a ripple you generate by reacting, remembering, and expecting.

Now, it's time to drop the final illusion:

You're not living *in* time. You are generating time.

Yes — you are the clock.

Time Doesn't Happen to You — It Happens *Through* You

Every moment you interpret an event, assign meaning, compare outcomes, regret the past, or anticipate the future — you're ticking.

Not on a wall. Not in space. But within your field of perception.

Your body may age by biology. But your experience of time is born from your ripple response.

- When you react, you delay.
- When you remember, you extend.
- When you judge, you fragment.
- When you wait, you stretch.

This is not metaphor. This is ripple mechanics.

You literally shape the time you experience — through your choices of attention, attachment, and alignment.

Time Is Not a Line — It's a Loop You Activate

We think of time as linear — birth to death, past to future. But in the UED framework, time is a looped consequence of ripple memory and unresolved alignment.

Every moment you hold onto something that's already passed, or try to force something that hasn't arrived — you spin the loop again. And that loop is what you call your "timeline."

But here's the punchline: That loop isn't running on its own. You are spinning it.

You are the loop-maker, the delay-giver, the attention-keeper. The universe doesn't demand this from you. It's your response to ripple that creates time as you know it.

If You're the Clock, You're Also the Key

This realization isn't meant to shock you. It's meant to free you.

Because if you are the generator of time experience, you can also be the one who pauses it. Who slows it. Who steps out of it. Who chooses when to tick — and when to simply be.

You are not a victim of time. You are the cause of its grip.

And in that, lies your greatest power.

Stillness Is Not the Absence of Time — It's Your True Self Without Ripples

You've felt it before:

- In deep meditation
- In nature, without your phone
- In moments of flow, where hours passed like seconds
- In moments of love, awe, surrender

What was gone in those moments? The clock in your mind. And what remained? You — untouched, alive, free.

Stillness doesn't erase the ripple. It just refuses to engage it. And in that refusal, the ticking ends.

Not because the universe stopped. But because you remembered who you are beneath time.

The Final Shift: From Time-Tracker to Presence-Carrier

You no longer need to track your age, your journey, your milestones, your redemption.

You only need to remember:

Time is not something you measure. It's something you create — when you forget you were never meant to measure.

So the question is no longer:

- *What time is it?*
- *How much time do I have?*
- *When will this end?*

The only question now is:

Do I need to tick right now — or can I just be?

Final Quote:

"You are not running out of time. You are running loops of attention. And the moment you stop spinning —You stop ticking. And there, you find yourself — Timeless, untouched, whole."

Closing Note: The Ripple is Eternal

In the beginning, we questioned what time is.

By the end, we've realized: Time is not a container. It is a consequence.

It is not something moving around you — It is something emerging from you, the observer. A ripple trail that arises when presence breaks into reaction. When coherence meets delay. When stillness is disturbed — even subtly.

But even deeper than time, deeper than consequence, lies something else:

A field that never began and will never end. A presence beyond motion. A silence that holds every sound, every memory, every tick.

This is Structured Vacuum Energy (SVE) — The unshaken background behind every ripple. The stillness that doesn't resist disturbance, but allows it. The infinite canvas where ripple plays, time arises, and consciousness remembers itself.

Intelligence, Consciousness, and the Ripple Field

You are not just a body moving through time. You are a field of intelligence — conscious, aware, and alive.

And what you call time... is your attempt to organize your dance with consequence.

But consciousness is not born from time. And intelligence does not need clocks to function. These emerge from the ripple, but they are not bound to it.

When your awareness returns to the field — When attention is no longer claimed by memory or prediction — You reconnect with what you were before the first tick: Pure coherence.

Time Is Temporary. Ripple Is Eternal.

Time comes and goes with your focus. But ripple — the dynamic pulse of existence — is eternal.

It is not bound to beginning or end. It simply responds. It flows through creation, intelligence, emotion, gravity, and growth — But it remains inseparable from the field it disturbs.

And in this way, the ripple is not your enemy. It is your reminder.

That you are still in motion, Still alive, Still able to align.

What Now?

You've reached the end of this book. But not the end of time. And certainly not the end of awareness.

What changes now is your relationship to time. To consequence. To ripple. To memory. To being.

You no longer need to be the one pulled by time's illusion. Because now you know where it came from.

Not from the sky. Not from a god. Not from physics alone.

But from you — The observer. The witness. The source. The ripple.

"Time fades. But the ripple remains — Not to trap you, But to teach you how to dance Without forgetting the stillness beneath your feet."

Appendices

Appendix A: The Time Paradox — Science Without the Observer

For centuries, science has treated time as an independent, measurable quantity — a dimension as real and objective as space. We've built clocks, equations, theories, and entire cosmologies on it.

But there's one profound flaw in this structure: We forgot the observer.

Introducing Chavan's Time Paradox

Chavan's Time Paradox exposes a central contradiction in modern physics:

"We use time to understand the universe — but exclude the one who experiences it."

Mainstream science assumes that time is absolute or at least structurally measurable (as in General Relativity or Quantum Time). Yet it builds models without integrating the one entity that makes time meaningful: the observer.

Here's the paradox, broken down:

- Physics treats time as something "out there" — a ticking backdrop.
- But human experience of time is entirely "in here" — felt through delay, anticipation, memory, and attention.
- Theories ignore perception — even though every measurement of time is observer-dependent.

So we end up with equations that work mechanically, but fail experientially.

Why Time Equations Collapse Without Perception

Consider any time-dependent formula in physics:

- Acceleration: $a=dv/dt$
- Energy: $E=h\nu$ is frequency (cycles per time)

- Relativity: $t'=t1-v^2/c^2$

These assume uniform, divisible, measurable time — typically tied to atomic clocks or spacetime geometry.

But remove the observer, and a deep question arises:

Who experiences the change?

Without perception:

- There is no confirmation that time is passing.
- There is no reference to measure "before" and "after."
- There is no validation of delay or sequence.

Time becomes a ghost variable — used in calculations, but never felt, witnessed, or anchored in conscious experience.

Thus, equations may still compute, but their meaning collapses — they no longer explain what it *feels like* to exist in time.

Time Is Not a Dimension — It's a Consequence

In Universal Energy Dynamics (UED), time is redefined not as a fixed dimension, but as a field response:

Time is born the moment a ripple misaligns and is perceived with delay.

This introduces ATT — Alignment Threshold Time — the smallest measurable unit of experienced time, defined not by seconds, but by how long it takes an observer to align with a ripple event.

- In perfect alignment (flow states, deep presence): ATT $\approx 0 \rightarrow$ Time disappears.
- In high misalignment (trauma, stress, confusion): ATT increases $\rightarrow$ Time stretches.

This explains:

- Why time feels slower when bored.
- Why time disappears during peak moments.
- Why children and elders perceive time differently — not because of age, but because of perceptual ripple density.

So, time is not what the clock says. Time is what your consciousness resolves after delay.

The Scientific Contradiction

Let's restate the contradiction at the heart of mainstream models:

Science wants to describe reality without consciousness — But reality only becomes meaningful through consciousness.

We define events by their timestamp, but only the observer:

- Registers the event
- Assigns sequence
- Experiences duration

To describe time without perception is like trying to measure taste without a tongue — The instrument can log data, but the experience is lost.

Hence, any model that explains the universe without the observer is fundamentally incomplete.

Tying It All Together

Chavan's Time Paradox urges us to realign science with reality:

- Equations must incorporate ripple delay — not just abstract time symbols.
- Experiments must factor in perception thresholds (ATT) and observer states.
- Theories must evolve from clock-based physics to consequence-aware field dynamics.

Because if time is born from ripple misalignment and felt through perceptual delay... Then the real key to time isn't the clock —It's you.

Final Thought:

"The paradox is not that we don't understand time. It's that we never included the one who does."

Appendix B: ATT and Ripple Mechanics Primer

If time is not a dimension, but a consequence, then we need a new way to understand how and when time appears. That's where UED's ripple mechanics come in — and with it, a revolutionary concept:

ATT — Alignment Threshold Time The time you actually experience is not what the clock tells you, but how long it takes for you to align with what is happening.

This appendix introduces the core terms and mechanics that replace traditional time-based physics with perception-aware dynamics grounded in the structure of ripple interaction.

Key Concepts in Ripple Mechanics

1. ATT (Alignment Threshold Time)

Definition: The minimum delay between an external ripple (event, interaction, change) and the observer's successful alignment with it.

In simpler terms, ATT is the delay you experience before you fully "feel" or "respond to" reality. It governs how fast or slow time feels — and it varies from moment to moment.

- Low ATT → Flow states, intense presence, timeless moments
- High ATT → Distraction, misalignment, trauma, stress, lag

UED proposes that all subjective time is a product of ATT, not seconds. Time = the lag of alignment, not the movement of a second hand.

2. Ripple Rhythm

All reality — events, thoughts, movements — is ripple-based. These ripples originate from misalignments in the stable field (SVE).

Ripple rhythm refers to:

- The frequency and harmony of ripple flow
- The coherence between internal ripple patterns (intent, emotion, attention) and external ones (environment, interaction, task)

When your internal ripple rhythm matches the outer ripple field, you don't notice time — because ATT is minimal.

When rhythms are misaligned, delay emerges — and time is felt.

3. Perception Time

Perception Time is not what clocks record. It is what the mind reconstructs based on:

- Delay (ATT)
- Emotional charge
- Ripple conflict or harmony

Two people in the same objective time can experience it completely differently — because they are perceiving through different alignment lags.

- 1 hour at a concert may feel like 5 minutes.
- 5 minutes in a dentist's chair may feel like an hour.

Why? Not because time bends — but because perception does, based on ripple tension.

4. Ripple Realignment vs. Ripple Residue

When a ripple enters your field, one of two things happens:

- You align with it immediately → no time is felt.
- You resist or delay → the ripple leaves a residue — stored memory,

emotional trace, interpretation.

This ripple residue is what constructs your timeline. It creates memory loops, future expectations, and a personal sense of past/future.

UED asserts:

- Time isn't just delay — it's your echo of that delay.
- Your timeline is not fixed — it's your accumulated ripple residue.

The less residue you carry, the more timeless your experience becomes.

Diagrams to Reframe Time

Diagram 1: How Time Arises

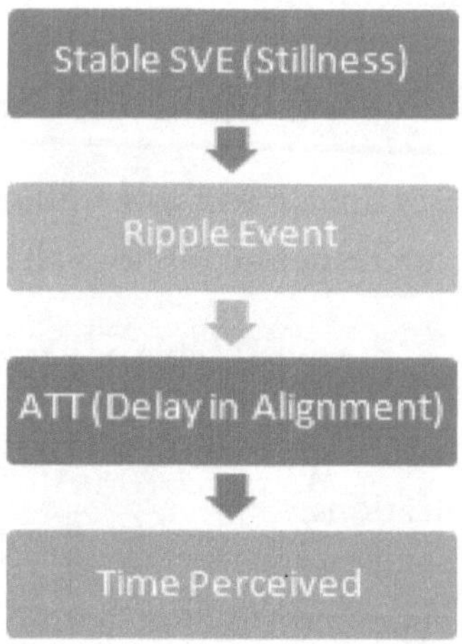

Time is not a precondition — it is the output of a ripple-delay sequence.

Diagram 2: Time as Constructed Memory

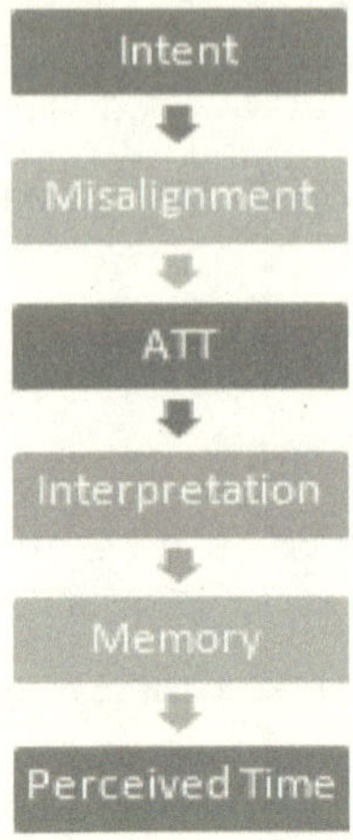

This shows how *subjective time* is generated not by clocks, but by your response to ripple misalignment.

UED Replaces Time With Alignment Lag

Traditional physics measures time in seconds, milliseconds, Planck units. But UED proposes a more accurate, field-conscious alternative:

Replace time t with ATT — the measurable or perceptible lag between event and observer alignment.

This shift has radical implications:

- In cosmology: Time "begins" not with a bang, but with the first ripple misalignment.
- In psychology: Emotional trauma is time stored in high-ATT ripple residue.
- In therapy: Healing is not about passing time, but realignment — reducing ATT.
- In consciousness: Enlightenment is ATT → 0 — response without ripple resistance.

UED doesn't reject time — it reveals what time actually is: A perceptual by-product of ripple delay in a conscious field.

Final Insight

You don't measure time. You feel delay. And when the delay disappears — so does time.

Appendix C: The Observer's Equation — You Are the Clock

At the heart of this book lies a radical insight:

You don't live in time. You generate it.

This is not poetic metaphor. It's a consequence of ripple dynamics and perception: Time is not what the universe gives you. Time is what you create when you interact with it.

This appendix unpacks how you — the observer — are the true clock, and how your attention, memory, and emotional response generate the time you feel.

The Personal Clock Is Made of Attention, Not Gears

Modern clocks run on mechanisms — gears, pulses, oscillations. But the clock inside you runs on something much subtler: awareness under tension.

Your personal clock doesn't tick evenly. It ticks in:

- Reactions
- Emotional charges
- Ripple residues
- Looping thoughts
- Delayed alignments

Each time you respond with misalignment, you store a delay — and that delay becomes felt time.

The Ingredients of Time Perception

Let's break down the components of the observer's clock:

1. Attention

Where your attention goes, energy flows — and so does perceived time. If your attention is fragmented or scattered, time feels slow or chaotic. If it's focused and fluid, time dissolves.

2. Reaction

Every time you react (instead of respond), you create ripple resistance. This resistance increases ATT — the perceived delay — which lengthens your experience of time.

3. Memory

Memory is ripple residue. It stores past delays and plays them back. You don't just remember events — you remember how long they felt. The more you loop on memory, the more time you create retroactively.

4. Emotion

Emotion is ripple turbulence. Strong emotions stretch ATT — grief, fear, shame create time loops. Peace and joy reduce ATT to near-zero — and time seems to vanish.

Together, these four components don't *track* time. They form time.

Your clock isn't on your wrist — it's in your alignment.

Inertia of Experience — Not Movement in Matter

In conventional physics, time is linked to movement — celestial motion, atomic oscillation, or entropy flow.

But UED reframes this entirely:

Time is not caused by movement. It is caused by misalignment between ripple and response.

That means time doesn't arise from things changing — It arises from your lag in catching up with change.

Time = Inertia in your experience — how long you take to fully absorb, respond, or align with a ripple.

- Matter may move, but unless you feel lag, no time is experienced.
- You could live an entire day in harmony and say, "Where did the time go?"
- Or spend five minutes in distress and feel like an hour passed.

Time is the inertia of misalignment.

Real-Life Ripple Examples

Let's decode common time experiences through the UED lens:

Trauma

- Ripple with intense misalignment
- High ATT — slow internal processing
- Time stretches and becomes sticky
- "It felt like forever," or "It's been years but I'm still there"

Joy or Flow

- Ripple matches your internal field
- ATT ≈ 0 — immediate coherence
- No delay, no residue
- "Time flew," or "I lost track of time"

Boredom

- Ripple rhythm is too slow or repetitive
- Your field seeks stimulation but gets none
- ATT remains high due to internal-external mismatch
- "Time is dragging," or "This is taking forever"

These examples show that time is not universal — It's not even personal in the emotional sense. It is perceptual, functional, and entirely tied to your ripple processing system.

Final Insight: The Observer's Equation

If we were to write a UED-based time equation for the individual, it would be:

Perceived Time = f(ATT) = Ripple Delay / Alignment Capability

Every second you feel is a result of:

- How misaligned you are
- How quickly you adapt
- How little resistance you apply

When alignment is perfect, time collapses into presence.

You Are the Clock — And That's Good News

Because if you are the source of time perception...You can also be the one who stops it, slows it, or chooses not to tick for a while.

You're not a slave to time. You are the rhythm behind it.

And when you realign — time doesn't end. It simply becomes irrelevant.

Appendix D: The Illusion of Beginning and End

One of the most powerful insights born from the UED framework is this:

"Beginning and End are not cosmic facts — they are psychological perceptions born from ripple memory."

We mark events with start points and finish lines — births, deaths, discoveries, collapses. But these are not truths of the universe. They are interpretations made by the observer, whose field memory gives rise to the illusion of linearity.

In reality, nothing truly begins. Nothing finally ends. What we call the "start" or "end" is simply a transition of ripple form.

Deconstructing the Big Bang and "End Times" Myths

Let's apply this to two central cosmological narratives:

1. The Big Bang Fallacy — A Beginning Without Tools to Measure It

Mainstream science frames the Big Bang as the beginning of time, space, energy, and the universe itself.

But this leads to a logical contradiction:

"If time began at the Big Bang, how can we talk about what happened *before* it — or even say it *happened* at all?"

UED resolves this paradox by revealing:

- There was no bang — only the first ripple.
- There was no explosion — only misalignment in the silent, stable field of Structured Vacuum Energy (SVE).
- Time did not exist *before* — because ripple had not yet introduced

delay.

So the Big Bang isn't a beginning. It is the first perceived consequence.

2. End Times — The Projected Closure of a Timeline

Religious and scientific theories alike predict a final end:

- Heat death
- The great crunch
- Apocalypse
- Cosmic silence

But these visions still rely on observer-based narration — They assume that the timeline must end because it began.

UED says otherwise:

- The field never ends — it only shifts alignment.
- What dies reforms.
- What concludes reorganizes.

There is no finality in ripple mechanics — only phase transitions.

Reframing Beginning and End

Let's reframe both concepts through ripple logic:

There Was No Beginning

- Only perfect stillness in the stable SVE field.
- Then, a misalignment — not an origin, but a deviation.
- Time began as a result, not as a precondition.
- The universe didn't start. It became.

There Will Be No End

- Realignment can occur. Silence may return.

- But the field remains, ready for the next ripple.
- What ends in one domain continues elsewhere in a transformed state.
- You don't leave the field — you reconfigure within it.

Philosophical Parallel: Becoming and Transformation

These insights are not exclusive to physics — they echo the deepest teachings of philosophy and mysticism:

"Nothing begins — it becomes." "Nothing ends — it transforms."

This is the universal rhythm:

- A drop becomes a wave.
- A wave becomes mist.
- Mist becomes cloud.
- But the water never disappears.

So it is with existence, consciousness, and the ripple we call time.

Diagram: *Observer Timeline vs. Field Continuity*

Here's how UED distinguishes subjective illusion from universal truth:

Observer's Timeline

Memory → Experience → Death (Closure)

Ripple Continuity

Field Tension (No Origin) → Alignment → Realignment Flow (No Finality)

The observer's timeline is a narrative. The field's rhythm is a flow.

Final Insight

You keep asking when the universe began. But that's a question born of ripple delay. The universe never started. It was always waiting for you to align.

Appendix E: Rewriting Time in Science — Experimental Implications

In the current scientific paradigm, "t" — time — is treated as an independent, absolute, or relativistically warped variable. It's assumed to be *measured* through clocks and interpreted through frameworks like spacetime geometry and thermodynamic entropy.

But if time is not an entity, and instead a consequence of ripple delay experienced by an observer, then science must undergo a serious paradigm shift.

This appendix offers a practical roadmap: How to test and implement time as ripple delay — not as dimensional flow — in real scientific studies.

UED-Based Experimental Proposals

1. ATT Mapping in High-Energy Systems

Hypothesis: Ripple delay (ATT) increases with complexity, field turbulence, or energy concentration.

Method:

- Use reaction-time probes embedded in energy-variable chambers.
- Track perceptual thresholds in human subjects or AI sensory systems.
- Compare ATT at rest vs under energy flux or EM stress.

Prediction: Increased energy leads to longer ATT, not faster perception, due to ripple noise and misalignment.

2. Time Distortion as Ripple Lag, Not Relativistic Dilation

Reframe Einsteinian time dilation as alignment lag due to ripple resistance in motion.

Test:

- High-speed particle observations (like muon decay): Instead of measuring decay rates purely by velocity, analyze field ripple density near particles.
- Use interferometers or ripple-simulation devices to track misalignment rate.

Goal: Replace relativistic time dilation interpretation with field misalignment delay.

3. Field Fatigue and Temporal Perception Studies

Objective: Test how ripple field saturation affects subjective time flow.

Method:

- Cognitive load experiments, long meditation sessions, or EM field immersion.
- Monitor subjects' perceived passage of time.

Prediction: High-ripple environments slow perception (increase ATT), while harmony-inducing environments reduce ATT and distort time perception toward timelessness.

Scientific Theory Implications

Relativity (GR & SR)

Current Model: Time bends in gravitational fields or stretches near light-speed motion.

UED Model: No need for spacetime curvature — ripple rhythm restriction explains the same phenomena.

- Gravity affects local field density and ripple rhythm, delaying realignment and thus *time perception*.
- Replace curvature with ripple distortion maps.

Quantum Decoherence

Mainstream View: Superpositions collapse due to entanglement and probabilistic wavefunction decay.

UED View: Decoherence occurs due to ripple misalignment thresholds being exceeded.

- Quantum fuzz is not temporal — it's field fluctuation awaiting ripple resonance.
- Introduce ATT thresholds in wavefunction models.

Gravity

Standard View: Gravity bends time and space.

UED Interpretation: Gravity is a function of ripple field tension.

- Time does not warp — ripple equilibrium shifts.
- Heavier mass = more ripple inertia = slower local ATT.

Roadmap: Replace "t" with ATT in Equations

To integrate UED into mainstream physics, we don't need to discard formulas — we redefine the terms.

Instead of:

$$v = d/t$$

We use:

$$v = d/ATT$$

And:

$$F = ma = m \times (\Delta v/ATT)$$

Where:

- ATT is the measured or simulated Alignment Threshold Time, not clock time.
- ATT can be calibrated through:
    - Reaction thresholds
    - Ripple latency scanners
    - Sensory alignment devices (AI-augmented or neurological)

This replaces ambiguous or absolute time with contextual alignment lag, measurable in both human and non-human systems.

Final Insight

"The future of physics lies not in conquering time — but in aligning with what causes it."

Appendix F: Spiritual Integration — The Timeless States

Throughout this book, we've dismantled time as an absolute force and reframed it as a consequence of ripple misalignment — a function of perceptual delay rather than a fundamental dimension. But this insight isn't just scientific. It has been hinted at — even *lived* — by mystics, sages, meditators, and everyday people for thousands of years.

Timelessness is not fiction. It is not death. It is a shift in your alignment with reality.

This appendix explores how spiritual traditions and altered states — including sleep — reveal the true nature of time as something you can exit, soften, or even dissolve, without leaving the body or the world.

1. Sleep — The Natural Exit from Time

Perhaps the most overlooked example of timelessness is one you experience every single night:

Sleep is the state in which the personal clock stops ticking.

- In deep, dreamless sleep, there is no "I" to register seconds, regrets, plans, or identity.
- You don't measure time during sleep — you *re-enter* it only when you wake.
- If time were an absolute container, you would feel it even while unconscious. But you don't.

This proves that:

- Time requires perception.
- And perception requires a ripple misalignment to track.

When you sleep, especially in non-REM or deep stages, the ripple field stabilizes, ATT approaches zero, and your personal experience of time ceases.

This is not just rest — it is field-level realignment. A biological demonstration of UED in action.

2. Meditation, Presence, and the Still Mind

All great spiritual paths teach some form of presence — and with it, they describe the loss of time:

- Zen: *No-mind, no-time.*
- Advaita: *The Self is beyond time.*
- Sufi Whirling: *Ecstatic timelessness.*
- Christian mystics: *The eternal now.*

In deep meditation:

- Internal ripples settle.
- The need to interpret, predict, or remember weakens.
- ATT shrinks — sometimes vanishing altogether.

You stop ticking. You just are.

Time doesn't disappear from the world. It disappears from your field — because there is no delay between reality and response.

3. Flow and Creativity — Time's Natural Collapse

Artists, athletes, and innovators often describe flow states where:

- Hours vanish.
- Effort feels effortless.
- Thoughts dissolve.

This is ripple resonance in real-time.

- Inner intent matches outer challenge.
- Misalignment is minimal.
- ATT ≈ 0.

Flow isn't mystical. It's field-synchronized living.

And the consequence of this coherence? Time fades. Because you're no longer generating it.

4. Mystical Liberation — The End of the Loop

Across cultures and ages, enlightened beings have described one thing in common:

Freedom from time.

Let's map this across traditions using UED interpretation:

***Moksha* (Hinduism)**

- Liberation from the cycle of birth and death (samsara).
- UED view: Detachment from ripple residue (karma).
- No longer reacting to consequences = exit from the ATT loop.

***Nirvana* (Buddhism)**

- Extinguishing the flame of desire.
- UED view: No longer generating ripple turbulence.
- Without ripple tension, time has no traction.

***The Kingdom Within* (Christian Mysticism)**

- Eternal life as a state of being, not afterlife.
- UED view: Living without delay — as pure alignment.
- Heaven is not a destination. It is ATT = 0 in your field.

These aren't fantasies. They are ancient languages trying to describe the timeless state experienced when ripple delay no longer defines the self.

Ripple to Silence — The Continuum of Time Exit

Let's organize levels of time detachment by ATT intensity:

State	ATT Level	Time Felt	UED Interpretation
Trauma / Anxiety	High	Time slows	Ripple overload, misalignment
Ordinary Wakefulness	Moderate	Linear time	Functional delay, partial resonance
Flow / Creative Absorption	Low	Time fades	Coherence between internal & external ripples
Deep Sleep / Meditation	Very Low	Time absent	Ripple field stabilized
Mystical Liberation	Zero	No time	Observer disengaged from all misalignment

Final Insight

Timelessness is not a mystery. It is your original condition.

Before memory, before delay, before identity — you were already there. What we call "spiritual awakening" is just the realization that you can stop generating time.

Time is your echo in a distorted field. When you return to stillness — the echo ends.

And what remains... is *you*.

Further Reading & References

Author's Research & Core UED Works

- **Chavan, Sandeep.**

Chavan's Time Paradox: Why Time Can't Exist Without an Observer (White Paper, 2025)

- **Chavan, Sandeep.**

The Illusion of Time: Causal Events and Energy Transitions (UED Book Series, Vol. 3)

- **Chavan, Sandeep.**

Structured Vacuum Energy: The True Background of Reality (UED Core Physics Reports, 2024–2025)

- **Chavan, Sandeep.**

Perception and the Ripple Field: Realigning Scientific Experience (UED Cognition Series, 2025)

- **Chavan, Sandeep.**

The Human Game: ATT, Memory, and Time Loops (Book 2 in The War of Thoughts series)

Comparative Spiritual & Philosophical Works

- **The Bhagavad Gita** – Especially Chapter 11 (*"Time I Am, Destroyer of Worlds"*) → Insight into divine time as consequence and dissolution.

- **The Dhammapada** – Teachings of Buddha on impermanence and liberation from desire (time loops).
- **The Tao Te Ching** by Lao Tzu – Timelessness through effortless alignment (Wu Wei).
- **St. Augustine, *Confessions*** – "What then is time? If no one asks me, I know."
- **Alan Watts, *The Wisdom of Insecurity*** – Exploration of timeless presence in human life.
- **Eckhart Tolle, *The Power of Now*** – The modern spiritual perspective on detaching from time.

Scientific and Psychological References

- **Einstein, Albert.**

Relativity: The Special and General Theory (Note: Referenced for contrast with UED model)

- **Carlo Rovelli.**

The Order of Time → Philosophical reflections on quantum time and relational physics.

- **Julian Barbour.**

The End of Time: The Next Revolution in Physics → Strong parallel: Argues that time is an illusion created by change.

- **David Eagleman.**

The Brain: The Story of You → For insights into how memory and perception create subjective time.

Cross-Disciplinary Explorations

- **Iain McGilchrist.**

The Master and His Emissary → Discusses how the divided brain processes time, experience, and wholeness.

- **Rupert Spira.**

Being Aware of Being Aware → Pure non-dual awareness without time.

- **Nassim Haramein**

Quantum Vacuum & Unified Field Theories → For those interested in fringe science echoing SVE models (read critically).

Closing Note:

Many of the works above represent **different worldviews** — materialist, mystical, experiential. This curated list is not to validate or disprove them, but to **show the landscape of thought** around time, perception, and the observer.

The goal of *TIME: The New Definition* is not to argue who is right, but to invite you to **feel what is true.**

Because **truth**, like time, must be **aligned with — not just calculated.**

Acknowledgments

This book is not just the result of writing — it is the result of a lifetime of **listening**.

Listening to silence. To doubt. To reality. And to every question that time ever asked me.

I owe deep gratitude to the **universe itself** — for revealing its intelligence not through complexity, but through patterns that whisper truth to those who are willing to pause and perceive.

To my **students** at the Gyrus Institute — Thank you for your curiosity, your confusion, and your courage. You reminded me that the real value of teaching lies not in answers, but in how deeply we learn to ask.

To my **family**, who silently supports the path I've chosen — Your quiet acceptance, even without full understanding, means more to me than any applause.

To **friends** and fellow thinkers, both known and unknown, who've debated, challenged, questioned, and inspired — your ripples have helped shape mine.

To those who read this not as a textbook, but as a mirror — Thank you. This work lives through your observation.

To **time itself**, Thank you for showing up just long enough for me to say: "I finally understand you — and now, I can let you go."

Lastly, to the **intelligence behind all things** — the structured field, the silent presence, the source of all ripples — I didn't invent this theory. I simply remembered it.

And now, I return it to the field — for others to find, feel, and possibly forget again, until it's needed.

With respect and resonance,

Sandeep Chavan

Independent Researcher, Observer, Ripple

About the Author

Sandeep Chavan is an independent researcher, educator, and author of over 100 works spanning physics, philosophy, human behavior, and metaphysical inquiry. Known for his pioneering framework Universal Energy Dynamics (UED), he challenges mainstream scientific assumptions by restoring the observer to the heart of reality.

With a background in engineering and decades of experience teaching physics to young minds, Sandeep blends rigorous logic with deep introspection. He is the founder of Gyrus Institute, where he encourages students not just to study science, but to question the foundation on which science stands.

His books are not academic exercises — they are invitations to see differently, feel deeply, and live intelligently. Whether rethinking gravity, decoding the soul, or redefining time, Sandeep's work consistently returns to one central truth:

"Reality is not something you pass through — it's something you participate in."

He lives simply, teaches passionately, and writes to awaken the silent intelligence within us all.

Don't miss out!

Visit the website below and you can sign up to receive emails whenever Sandeep Chavan publishes a new book. There's no charge and no obligation.

https://books2read.com/r/B-A-EPGPC-BYRJG

BOOKS 2 READ

Connecting independent readers to independent writers.

Did you love *TIME – The New Definition*? Then you should read *Feeling Like a Physicist*[1] by Sandeep Chavan!

[2]

Feeling Like a Physicist isn't about equations. It's about what happens when you sit long enough with reality to hear it whisper back.

This is not a science book in the conventional sense. It's a **personal record of consequence** — a journey through physics, philosophy, cosmology, and human behavior, driven not by academic pursuit but by a quiet need to make sense of the world without lying to myself. I didn't set out to prove anything. I set out to feel what was real, beneath what we're taught to memorize.

The result is **Universal Energy Dynamics (UED)** — a ripple-based way of understanding existence. Through this lens, force becomes resistance. Time becomes delay. Motion becomes field realignment. Even emotion, fear,

1. https://books2read.com/u/bPjxAY

2. https://books2read.com/u/bPjxAY

desire — they all emerge as ripple behaviors, not mental constructs. You'll find redefinitions of gravity, space, light, karma, the soul, and even God — not through belief, but through **field alignment and consequence logic**.

This book was written without a lab, without a label, and without any obligation to follow accepted frameworks. That's why it's honest. Not perfect — but honest. And I believe that matters more.

What you'll find here are **25 small chapters** — each one a moment of realization. Together, they map a path from questioning to clarity. From silence to structure. From feeling lost in complexity... to realizing that nothing is complex when viewed from the center of alignment.

I don't want followers. I want **feelers** — people who sense before they agree. People who are willing to sit with discomfort, to doubt what they've been told, and to find their own alignment through their own ripple.

This book is for the **quiet thinkers**, the ones who never needed validation to ask deeper questions. It's for the teacher who doesn't believe the syllabus anymore. The student who senses something's missing. The scientist who's tired of math without meaning. The soul that wants clarity without mysticism.

If you've ever felt that reality must be simpler — not smaller, but **more coherent** — this book is for you.

I didn't write it to teach you. I wrote it to **stand beside you**.

— Sandeep

Read more at https://www.sandeepjchavan.com/.

Also by Sandeep Chavan

Chavan's UED research

CMGED Research Report

Chavan's Cosmic Continuum (C^3)

Frequency

Ripples of Reality

Reality Realigned

The Great Realignment

I – The Cosmic Leverage

Chavan's Cosmic Theory of Relativity (CTR)

Chavan's Theory of Light

Feeling Like a Physicist

TIME – The New Definition

Echoes of War: The Russia-Ukraine Conflict and Its Global Ripples

Shattered Lives

The War Economy

Borders Redrawn

Arms and Allies

Collateral Damage

Justice in the Ashes

Somewhere in the Darkness of the Cosmos

The Blueprint of the Universe
The Cosmic Breath
From the Edge of the Universe

The Eternal Sage Chronicles
The Sage and the Sovereign
The Family Pillar
The People's Flame
The Warrior's Soul
The Bridging Mentor
The Guiding Light

The Exodus
The Last Cycle: 2075

The Fundamental Theory of Life & Existence
Energy in Action: The Soul Demystified
The Active Mind
Energy in Connection
Death: The Disconnection
Subconsciousness: The Chosen Disconnection
Desires: The Engine of Life

The Restoration of Human Knowledge
Restoring Bhagavad Gita

The War of Thoughts

Cracks in the Known
The Human Game
The Divine Dilemma
The Original Intent
The Answers

Ultra Modern Life Philosophy
The Duality of Laughter and Tears
Take It Easy
Compare to Accept

Standalone
It's Not AI, It's AHI - Amplified Human Intelligence
The Decision Paradox
The Four Sapiens
Malicious Script of Indian Polity
The IIT Legacy & Global Impact
The IIT Legacy & Global Impact
You are just a Version of Your Original
Win the Game You Didn't Choose
Always Keep Your Bags Packed
Reveal to Shield - Learn the Game of Facades
Engage Beyond Elections
Culture, Identity & Change: The Evolution of Indian Society
The Triangular Dynamics
Never Trust Without Doubt
If You Can't Change the Country, Leave the Country
Live Fearless with Fear
India: The Investment Magnet
Realignment
Powering a Billion Dreams
Universal Skill

Extension: From Brush to Algorithm
Human Side of Einstein
The Future Knows No Lines
Let's Bring the Generations to the Table
Unified Energy Dynamics
A Beat That Never Dies
Strategic Horizons
Right From Scratch
The Forgotten Innovators
Physics Fixing
The Savior Illusion
The Selfish Side
The Age of the Weakest Sapiens
Neural Wars
The Absorption: There Is No Death
Strange Therapist
Bubbles: Become a Drop
Beyond the Blue
Half-Scholar
The Law of Truth
Failure
Philosophy 2.0: A Structural Manifesto
Chaos: A Gentle Guide to the Unresolvable
Self-Help: New Syllabus
Rippling
Game: How to Play the Share Market
Out of Syllabus
Saffronized: The Code Behind the Crown
The Anger

Watch for more at https://www.sandeepjchavan.com/.

www.ingramcontent.com/pod-product-compliance
Lightning Source LLC
LaVergne TN
LVHW090309210926
848480LV00001B/185
* 9 7 9 8 2 3 0 7 7 1 7 0 8 *